PHOTOGRAPHER, PARATROOPER, POW

PHOTOGRAPHER, PARATROOPER, POW

A Wyoming Cowboy in Hitler's Germany

M Carroll

ANAMCARA PRESS LLC

For all those who have served, and those who love them.

ANAMCARA PRESS
P.O. Box 442072
Lawrence, KS 66044
www.anamcara-press.com

Copyright © 2016 Anamcara Press LLC
Copyright © 2016 M Carroll

Printed in the United States of America

ISBN: 978-1-941237-08-3

Excerpt of "Montevideo Maru" from the website www.montevideomaru.info, by Rod Miller. Reprinted by permission of Rod Miller.
Excerpt of "Sons of Dewitt Colony" from *Sons of Dewitt Colony* by Wallace L. McKeehan, editor. Reprinted by permission of Wallace L. McKeehan

All rights reserved. No part of this publication may be reproduced, stored in a retrieval system, or transmitted in any form or by any mechanical recording or otherwise, without prior written permission of both the copyright owner and the above publisher of this book.

CONTENTS

Foreword		ix
Preface		xi
Acknowledgments		xiii
Chapter 1:	The Forgotten Photo Album	3
Chapter 2:	Flatlanders	9
	Spring 1931	11
Chapter 3:	Western Romeo & Juliet	15
	Summer 1931	17
Chapter 4:	Reprobates & Malcontents	21
	Fall 1931	26
Chapter 5:	The Incursion Of The Clans	29
Chapter 6:	Tea-Time On The Frontier	33
Chapter 7:	The Republic Of Texas	37
	Spring 1932	40
Chapter 8:	Regulators, Moderators, Outlaws and Other Heroes	45
	Summer 1932	49
Chapter 9:	Antebellum Cowboy	57
	Spring 1933	62
Chapter 10:	The Gonzales Invincibles Company	67
	Summer 1934	71
Chapter 11:	Mavericks	77
	Summer 1935	82
Chapter 12:	Trailblazers	87
	Summer 1936	95
Chapter 13:	Code Of The Cow Country	101
	Fall 1936	106
Chapter 14:	Just Married	109
	Winter 1936	125
Chapter 15:	Emigrants & Uprisings	131
	Summer 1937	137
Chapter 16:	Miners, Railroaders, & Ancient Hybernians	143
		143
	Winter 1938	147
Chapter 17:	Little Bighorn	151
	Medicine Wheel	155

Chapter 18:	Cattlemen, Grangers, And Range Wars	159
	Fall 1939	166
Chapter 19:	Winds Of Change	169
Chapter 20:	A Born Paratrooper	177
	Fall 1940	185
Chapter 21:	"90 Day Wonders"	189
	Summer 1941	193
Chapter 22:	Photo Lab Non-Com	197
	Winter 1941	202
Chapter 23:	Skymen & Submarine Warriors	207
	Summer 1942	217
Chapter 24:	Love And War	221
	Fall 1943	237
Chapter 25:	Photographs, Faux Pas, and Fond Regards	241
	March 1944	246
Chapter 26:	Tent City	249
	April 1944	252
Chapter 27:	Booze And Brawls	255
Chapter 28:	Pig And Whistle	263
	May 1944	269
Chapter 29:	The Last Photo Op	273
Chapter 30:	Signals In The Dark	279
Chapter 31:	Screaming Eagles	283
	8 June 1944	287
Chapter 32:	If At First You Don't Succeed, Escape, Escape Again…	289
	13 June 1944	291
Chapter 33:	Starvation Hill	297
	July 1944	299
Chapter 34:	Trauma Of Capture	301
	Later In July 1944	305
Chapter 35:	The Escape Bug	307
	August 1944	310
Chapter 36:	Oflag 64	313
	October 1944	317
Chapter 37:	No Escape	329
Chapter 38:	Letter From Home	335
	January 1945, New York City	337

Chapter 39:	The Fifth And Final Escape, Sort Of	341
	January 1945, Poland	344
Chapter 40:	Rembertow	349
Chapter 40:	Escape From Odessa	353
Chapter 40:	Homecoming	359
	Spring 1945	361
Chapter 40:	Zippos And Hosiery	367
	Sometime After The War	370
Chapter 40:	Missing In Action	373
	Spring 2014	376
Appendix A	The Tale of Abraham's Ghost	383
Appendix B	The Jacob Carroll Migration	385
Appendix C	The History of Rowan County, NC	387
Appendix D	Peter Carroll	389
Appendix E	George G. Carroll	391
Appendix F	Coffin Family Papers 1864-1977	393
Appendix G	Standard Operating Procedure for the Jump	395
Appendix H	Descendancy Chart Peter Carroll & GG Carroll	399
Appendix I	Geroge W. Carroll Family Tree	400
Appendix J	George Granville Carroll Family Tree	401
Appendix K	Peter Carroll Family Tree	402
Appendix L	John M. Kirby Family Tree	403
Notes		404
Bibliography		426

FOREWORD

As the daughter of a World War II Darby's Ranger, I have read, researched and written much about warriors of the Greatest Generation. I know how much work goes into even the shortest of combat articles.

Therefore, I was astonished at the depth and breadth of the historical research contained in A Wyoming Cowboy in Hitler's Germany, as well as the author's ability to transmit it to the reader in an interesting and thoroughly enjoyable fashion.

Through the adroit use of photos, letters, documents and prose, the author brings to life — from boyhood through World War II — three All-American brothers, each of whom worked his way up to the rank of officer. All three left for war: a commander/paraskier, a submarine navigator and a photographer/paratrooper.

The two paratroopers were reported missing in action during the Normandy D-Day invasion. One of them was later documented to be a German prisoner of war. Of the three brothers who marched off to battle, only two came home.

This book relates a compelling story, skillfully written. It is a must read for those who love history and especially those fascinated by World War II history.

Marsha Henry Goff, author of *A Crazy Plan: Darby's Rangers Historic Last Stand at Cisterna*

PREFACE

In these pages, you'll find facts and memories about a family. It is not easy to find the beginning or the end of a family and so I selected a specific topic to explore. That topic is the mystery surrounding Robert C. Carroll who was the 1st Commanding officer of the 501st Parachute Infantry Regiment of the 101st Airborne during WWII. I've also tried to explore the history that created men like Bob, and his brothers Arthur and Harold, by looking back to their great-grandfathers George Washington Carroll and James McCarroll. Additionally, the goal of this manuscript is to compile Harold Carroll's photographs, correspondence, memoirs, and other memorabilia regarding Bob's, Arthur's and Harold's service during WWII for posterity.

I have drawn from many original and unpublished sources in the creation of this manuscript, as well as internet sources. You can look on me as an ambitious scrap-booker. I have compiled snippets of gossip, newspaper articles, pictures, military records, and memories to weave a story.

Stories are central to family memories and to historic memory in general. The stories in these pages have come from the mouths of family members in addition to more historically verifiable sources. To avoid the natural confusion that might occur deciphering fact from conjecture, I have separated stories where there is little or no corroboration from an outside source under the designation of the year they took place. Most passages described in the year they occur are based on family stories, often verified by more than one family member; some were drawn whole-cloth to illustrate a person, place or time. I have recorded events in chronological order for reading ease.

M Carroll

ACKNOWLEDGEMENTS

This book would not exist were it not for the encouragement, patience and input of Patricia Carroll.

I owe much thanks and appreciation also to Emmanuel Alain, Museum St. Mere Eglese, France, for reviewing source material; John Bodle for his patience and support; Mildred Carroll and family for her candid interview and photos; Bill and Nelda Carroll and family for a warm welcome and sharing stories; Rosalee Ruth Carroll for sharing the Carroll Memory Book; Lilas Cox for sharing family photos and stories; Lavina Fuller for sharing stories and a chance to view Peter Carroll's cabin site; Michael Gormally, former head of the VFW, for connecting me to the Eisenhower Museum; the Eisenhower Museum and staff of Abilene, Kansas, particularly Dan Holt, Kathy Strauss, and Herb Pankratz for reviewing the source materials and reproducing some twenty photographs; Rachel McLaughlin for her heroic efforts teaching me to use Adobe Creative Suites software; Vicki Julian for her encouragement and editing (although no book of this length and complexity can be completely free from errors, there would be many more if not for Vicki); and thanks also to the many others who listened and put up with me over the years it took to complete this work.

Ó Cearbaill

Carroll

PART I

Snapshots left at county chemist reveal treasure tro[ve]

The forgotten pho[tos]

MAE WESTS? Life-jackets appear to be worn here but they were issued to be worn only in aircraft when flying over the sea, in case of ditching. They were left in the aircraft once the order to "break up" was given.

STRIPES: Several pictures of Dakota (C47) aircraft were taken at Ashwell. Some show D-Day stripes, some don't. Were these snaps taken on the same day? The white bands show that some photos were taken on, or after, D-Day.

AIRCREW: Thought to be B Company paratroopers during a training exercise at Ashwell Airfield, Rutland. Ashwell was the parachute school for the division.

PHOTOGRAPHER? Platoon leader from B Company, Second Lieutenant Stanley Weinberg from Teaneck, New Jersey – thought to have taken the photos.

by Felicity Newson

A CHANCE discovery of old photographs has revealed a nostalgic time capsule from the war era.

An album full of atmospheric pictures lay gathering dust for almost half a century after being claimed from a chemist's shop in Quorn.

Subsequent detective work has shown that the snapshots were taken by an unknown officer of the 505th Parachute Infantry Regiment, which had been changed in Wood Lane in the village.

The officer took the films to Mr William Green's chemists shop for

Story of war era revealed

parts of the story told in the photographs.

According to Mr Wills, research specialist in the activities of the 82nd Airborne, the pictures were taken on a Kodak Bantam camera, which was typically family type camera of the day.

THERE clues lie as to the treasures covered by the collection

First Battalion and its Co[mman]der, and the offic[ers] [in] dress was probably C[olonel] E. Ekman.

Some of the pictur[es taken] at the Ashwell airfield [show] was the site of the p[arachute] also houses specialis[t] in Infant Pathfinder [...]

One factor which h[elps] hake is the absence o[f] the photographs. The[re are] two Griswold bags va[...] curtains, sub machine [guns] rifles were carried wi[th them]

ANOTHER unresol[ved] over the Mae We[sts] [...] jackets which ap[pear] mentioned by one pict[ure]

Mr. Green and Mr [...]

CHAPTER 1

THE FORGOTTEN PHOTO ALBUM

The boxes of WWII memorabilia sat in my closet for years. Occasionally I would open them and sort through the musty, time-forgotten photos and scraps of paper. As I packed for a working trip to England, wondering what to wear in autumn, I remembered a newspaper clipping from Leicester, among the scraps. Snapping my luggage shut, I quickly searched for the clipping to take with me, and reread it on the way to the airport.

From the Leicester Mercury, May 29, 1989:

"A chance discovery of old photographs has revealed a nostalgic time capsule from the war era.

An album full of atmospheric pictures lay gathering dust for almost a half a century after failing to be claimed from a chemist's shop in Quorn.

Subsequent detective work has shown that the snapshots were taken by an unknown officer of the 505th Parachute Infantry Regiment, which had been charged in Wood Lane in the Village.

The officer took the film to Mr. William Green's chemists shop for processing—but never called back for the developed pictures.

And it was only recently, when Mr. Green's son Peter, was going through some of his late parents' possessions that the album was found and research into its background began.

With the help of Picture researcher Mr. Deryk Wills, Mr. Green managed to piece together parts of the story told in the photographs.

According to Mr. Wills, whose research specialty is the activities of the 82nd airborne, the pictures were taken on a 4X6 size folding camera, which was a popular family type camera of the day.

There is no clue as to the time span covered by the collection, some of the C47 aircraft have white bands painted on the fuselage and wings, while others do not.

The white bands show that theses photographs were taken on or after D-Day, because they were only painted on a few hours before the invasion started.

The location of the parade of paratroopers has been identified as the First Battalion area of the Quorn Camp, and the officer addressing them was probably Colonel William E. Ekman…

One factor which has puzzled Mr. Wills is the absence of weapons in the photographs. There are one or two Griswald bags visible in which carbines with machine guns and rifles were carried when parachuting.

Another question mark is over the Mae Wests, the life jackets which appear to be worn in only one picture.

Mr. Green and Mr. Wills were helped in their quest for information by Mr. Harry Montgomery of the White Horse Inn in Quorn, which is a popular haunt of ex-American servicemen paying return visits to their wartime base."

And so began my search for the photos my dad took just before D-Day that had been mysteriously rediscovered some 55 years later by a grieving Englishman.

The clipping, with its faded photographs, traveled with me as I explored the modern cities and quaint countryside of lovely England. I finished work on Friday in Nottingham and had the weekend off to explore, although I had to travel to the coast for work on Monday.

Unfortunately, I could find no way to get to Quorn, the place where the pictures were taken. Except for the revived old steam engines that recently reclaimed the route from Loughbourogh to Quorn for tourists on summer weekends, no rail trains passed through the region. And so I had little hope of visiting the tiny village on this brisk November day and instead focused my attention on the larger city of Leicester and the Mercury newspaper hoping to find a reproduction of the 1989 clipping so I would have better copies of the pictures of "Tent City" taken by Harold James Carroll, my dad. The pictures had been taken on his personal Brownie, he told me, and the film was dropped off at a local drug store just prior to D-Day. He never returned to pick up the developed photos.

The taxi driver stepped out to gather my bags. His name was Colin, and he was of Middle-Eastern, not Irish descent, reflecting a much smaller and more inclusive world than when my Dad was stationed here. He was there to take me to the rail station, but upon hearing about my search for my Dad's pictures he seemed as intrigued as I was. He offered to drive me to Leicester through Quorn for a sum I was willing to pay.

Quorn is about sixty miles south of Nottingham, a scenic drive through the heart of old England. Our first stop was the local Chemist shop (drug store). I knew only that the owner had been a Mr. William Green and his son Peter had published the photographs. We stopped at the only drug store in the village. The bell chimed as we entered. The owner of thirty years, an elderly gentleman with frizzy white hair wearing a forest green sweater, stood behind a broad counter at the top of a short set of stairs so that he oversaw the little shop. A woman at the cash register was ringing up a customer, and so I directed my questions to the pharmacist, although he was not particularly amiable to conversation. He said he'd owned the store for most of his working life, lived in the village as well, and no, there were no other Chemist shops in town, just this store. He'd never heard of Mr. Green, William or Peter, or my photographs. He said he couldn't help me.

Exiting into the rain I held little hope of finding my father's pictures. Colin had already informed me that the Leicester Mercury, although almost directly across from my hotel, would be closed until Monday, and my travel plans had me on a train for the coast Sunday afternoon.

I trudged up the wet cobblestone walkway toward the corner pub feeling dejected with Colin following along behind me. "Buy you a drink?" I offered.

The White Horse Inn sat low to the ground, its bleached stucco walls supported by ancient black beams. The area didn't bring in many outsiders except WWII vets and those interested in fox. Quorn had gained recent notoriety over controversy regarding the Hunt. Sixty years ago, in 1944, it was teeming with American troops preparing for the Invasion.

We stepped down into the dimly lit, low ceilinged drinking establishment. A warm fire crackled invitingly. The White Horse Inn exuded a timeless charm. A lovely young blond woman was tending to customers sitting along the old polished-wood bar.

Colin egged me on. "Why don't you go ask the bartender?" he prodded.

Why not? I'd come this far. I approached her with little hope, the weathered copy of the Leicester Mercury in hand. "Hello. I'm looking for my dad's pictures." I gave her a bit of an explanation, and before I was finished she pointed me toward the owner of the pub—no longer Mr. Harry Montgomery but a Mr. Ian White.

Ian stood in animated conversation with several regulars, hoisting a mug. Interrupting an ancient tradition, I approached him with my paper and my story, the taxi driver, Colin, encouraging me quietly from the sidelines.

Ian looked a bit surprised, and as I talked, he grabbed my elbow and steered me around one of the weathered old beams supporting the pub and pointed to a large, wood frame centered on the white stucco wall. Covered in glass, nicely matted and carefully arranged were my dad's photographs. Ian silently steered me toward another wall and another frame holding still more of dad's photographs.

He found the photographs in a box upstairs in the attic of the White Horse Inn in April, 2004 after buying the pub. The pictures impressed him, and because of the upcoming anniversary and the return of some of the vets to the area, he'd had the pictures professionally matted and framed.

I was dumbfounded. I couldn't believe my luck finding the original pictures so carefully preserved. There he was. One eyebrow raised, the sparkle that belies the frown, Harold James Carroll gazed down at me from the historic wall of the White Horse Inn, Quorn, England.

Art, Harold, & Bob Carroll, Sheridan, 1921

CHAPTER 2

FLATLANDERS
If you aren't from here, you're a flatlander.

Bob, Art and Harold grew up in the shadow of purple mountains listening to the wind in the pines, and real and imagined tales of a turbulent past. The Great Depression extended to the smallest town and valley and the boys had a steady, if poor, upbringing. Their father, Arthur, a miner in his youth, worked hard on the Burlington railroad, and their mother, Myrtle, worked hard at home. They grew up three to a bed as youngsters. This probably helped them to stay warm during the bitter cold spells that blew down from Canada and over the Bighorn Mountains into Sheridan, Wyoming.

They lived on the edge of town when they were young and Harold learned to ride horses from the neighboring Crow Indian children he played with. They rode bareback and explored the foothills and nearby mountains. Although they lived outside of town, Crow, Arapaho, Cheyenne, and other Indians could be seen in town during the day, but they were mostly barred from access to the stores that lined the main street boasting modern goods and contraptions from back East.

It was 1931 and times were hard. The boys all contributed to the family income. By the age of eleven Harold did his part by bringing home his earnings as a newspaper boy, waving to the old folks rocking on their porch swings. His granddad George and uncle Charley played checkers as Harold sped by on his bicycle, tossing commentary and comics onto front porches and under neatly trimmed bushes in the sweltering summer heat.

In the news in 1931: "The Star Spangled Banner" became the official national anthem, the Empire State Building opened for the first time in New York City, and Al Capone was tried and sentenced to 11 years in prison. Hitler, the appointed leader of the Nazi Party, with the backing of his storm troopers, challenged Paul von Hindenburg for the presidency of Germany.* Of more interest to Harold, the first Donald Duck cartoon played in local theaters and Babe Ruth hit his 600th home run.

The now popular automobile had leveled the tourism playing field and both the wealthy and those not so wealthy ventured out in their cars to see the grand sights and scenic places—places sacred to the Indians like Niagara Falls, and for the adventuresome, Yellowstone.

The Bighorn Mountains and the Wind River valley lie between Sheridan and Yellowstone, a bourgeoning tourist area in 1931. The pass over the Bighorns from Sheridan to Greybull on Highway 14, Granite Pass, rises 8950 feet over Shell Canyon with jagged cliffs creating a breathtaking landscape of sheer walls and deep gorges filled with rushing white water and lined with lush forests.

Bob, Art and Harold could tell the flatlanders by their trousers and their boots. They looked like city slickers. Some of them tried to sport western wear, but they were exposed by their walk and facial expression —an expression of pain, due to the experience their feet were having in their new boots, and the walk of one hobbled.

Myrtle, Bob, Harold, & Art, about 1926

---

* The Nazi Party began as the NSDAP, or Nationalsozialistische Deutsche Arbeiterpartei – National Socialist German Workers' Party.

SPRING 1931

Imagine how some folks would stop and survey the mountain, looking at the narrow rutted road that crawled up its side, snaking in cutbacks up, up, up. "What's the road like on the other side?" they'd ask, or, "Any other route to Yellowstone?" Some just turned and left.

Harold watched the big man pull himself out of the 1930 Model T, his bulk well over six feet, and lumber over to Bob who was standing in the sun talking to the owner of the gas station..

Bob straightened his lanky form and looked the man eye to eye. They shook hands.

Harold had come knocking on his brothers' bedroom door unannounced, one day not that long ago, and found Bob hanging by his fingertips from the door jam.

"What are ya doing?" Harold asked. Bob ran his feet up the wall, slipped his legs through his arms, and did a little flip, landing neatly on his feet in front of a smiling and adoring little brother.

"Stretching," he answered. "It makes me taller."

And Bob was tall, but the man with the Model T was bigger, though not in a muscular way. The man's face slackened at the jaw line, creating a bull-dog look. Jowly, thought Harold. Bob was strong, if a touch skinny, and just sixteen with bulging muscles and angular features. His hair was so light a shade of blond it was almost white, and his eyes could change from sky-blue and friendly to cool slate in the matter of seconds as they bore into you. He won any staring competition when anyone dared to challenge him.

Bob and the flatlander both loomed over Harold, a slight eleven year old. "This boy gonna follow in that truck?" the flatlander asked incredulous, his jowls wiggling slightly with each word. He wore a heavy overcoat although it was not so cold; only a little snow remained on the edges of the road and under the fragrant pines. Higher up the mountain there would be snow and ice. This is what the big man feared, and his shiny polished shoes, along with the shine on his forehead, revealed that he'd spent little time out of doors in the weather.

Harold stepped up to the man. "Yes, sir," he assured. "I'll follow you to the other side. I've done it plenty of times." A slight boy with a big grin, Harold was spunky. Blond hair was neatly combed and kept in place with oil. His shirt was tucked into his worn, though clean and neat pants and his boots were weathered. Ankle high and sturdy, they'd logged some miles up and down these hills.

Art usually helped with the flatlanders, but Art was occupied studying for his big exam coming up next week. Granddad George and Grandma Ella had him corralled at their place where they "supported him in his educational pursuits," as their mother, Myrtle, put it. Sometimes against his will, Harold imagined.

George and Ella had taken Arthur under their wing when he was just a little tike. Myrtle recently complained, "Why do you only take Art to your house for visits? Bob and Harold feel left out, Mother." Myrtle questioned Ella forthrightly over the pastry dough they were molding.

Ella sucked in her breath. It was hard to say. "Art is different." The middle child at fourteen, Art even looked different than Bob and Harold, favoring his mother's side of the family with brown hair and a slightly darker complexion. Ella had always thought he looked like her brother Charlie Kirby.

"Art likes to read, and pick apples; he helps Granddad George with chores, and he'll fetch things for me. He always likes what we have for supper… Bob and Harold prefer to play outside and I'm always nervous and watching for breaking tree limbs the way they like to climb! They just cause me such worry." Ella's honest answer stopped Myrtle in her tracks. She loved her boys equally, but they were quite different as each child had his own qualities after all.

Harold was Bob's little replica, although he had a way to go in size. Bob was his idol and he looked up to his older brother as to a hero. He worshiped the ground Bob walked on and literally followed in his footsteps — hiking, climbing, rappelling and now, to help get this flatlander across the pass and earn a few bucks. He followed closely behind in Bob's tire tracks through the snow and icy patches and around rocks and ruts and over blind rises.

They made it safely over the mountains and deposited Wiggly Jowls at Greybull, a bit paler but no worse for wear, the boys a bit wealthier and just a bit more wear on the tires.

"Can we spend some of the money at the candy counter?" Harold asked his big brother hopefully.

Bob grinned. "Let's go get Art. I'll bet he'd like to come along." After the return trip, all three boys crowded into the pickup, laughing and jostling for position. Since Indians weren't allowed in the store, Harold bought a piece of licorice for Thomas, his Crow buddy.

They walked down Marion Street together, where Granddad George Carroll had camped his first night in Sheridan before it was even a town. Bob tried to imagine what it was like back in his grandfather's time, when the sidewalks were wooden and the streets mud. The sun was setting and their elongated shadows led their feet down a timeless road.

Bob, Art, & Harold, about 1925

Myrtle & Arthur Carroll, Sheridan, 1914

CHAPTER 3

WESTERN ROMEO & JULIET
Do you bite your thumb at us, sir? ~ Shakespeare

Bob's, Art's and Harold's grandparents were, at best, on the opposite sides of a very barbed fence and, at worst, arch-enemies. Arthur Carroll's unlikely marriage to tall and winsome Myrtle Carroll, George Carroll's eldest daughter, must have raised some eyebrows.

Myrtle was born in Dayton, Wyoming in the foothills of the Bighorn Mountains. She had a big brother named Granville and three little sisters: Orpha, Ednis and Jewel. Myrtle was tall and slender, with a sly smile and sharp wit. And she fell in love with handsome Arthur Carroll of neighboring Story, Wyoming.

Arthur also had a big brother, Robert, and was one of five children, too — Myrtle, his older sister (not to be confused with his wife), and younger siblings Ruth and Jim. Arthur was the second eldest son. He was educated and skilled, and had worked in the mines since childhood as did most of the men of the family, including his father Peter, who was a miner and immigrant from the British Isles.

In contrast, Myrtles' parents, George and Ella Carroll, had roots in the old South and a moneyed background. And they were longer established in the Sheridan area than their neighbors, Peter and Martha Carroll.

George was a cattleman and rancher, Peter was a miner and farmer. George and Ella had moved north to Wyoming from Texas in 1881, leaving memories of plantation life behind. Peter and Martha had moved west to

Wyoming in 1898 from Iowa and homesteaded in the Bighorn Mountains. Peter and his sons worked as miners.

George was a Democrat, Peter a Republican. George was anti-union, Peter pro-union. The George Carroll family considered themselves English, or maybe Scots-Irish. Peter was Irish, although born in Scotland.

The two families had nothing in common except their last name (they were no relation), but their children, Arthur and Myrtle, would bridge the gap.

Imagine Arthur and Myrtle as the turn-of-the-20th-century version of Romeo and Juliet, the miner's son and the rancher's daughter, the shanty Irish and the lace curtain Irish. Married in spite of it all, they produced three children, all boys: Robert Collin, (Bob), born June 19, 1915; Arthur George (Art), born September 14, 1917; and Harold James (H), born June 29, 1920. All three boys went to war. Only two came home.

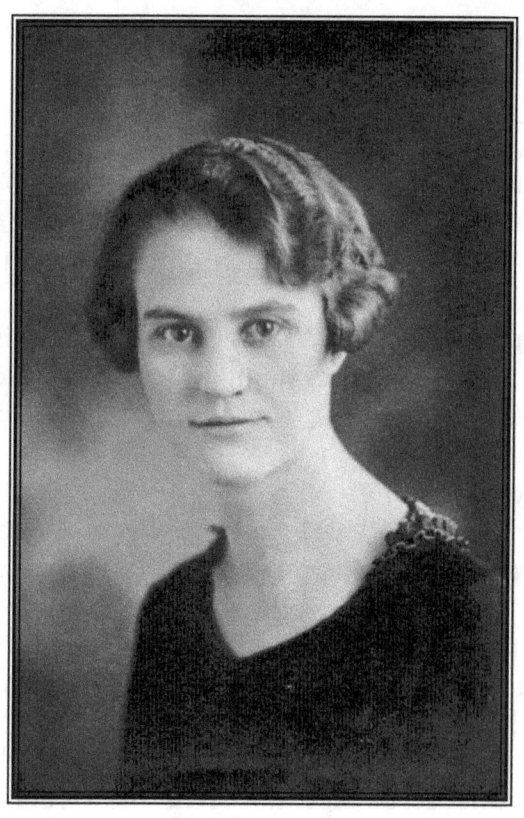

Myrtle Carroll, about 1910

SUMMER 1931

Bob, Art, and Harold, their bodies surprisingly still except the occasional butt-shifting necessary to keep numbness at bay, leaned in around the campfire and listened attentively. Granddad Carroll and Uncle Charley Kirby and other old timers began the story-telling as soon as the sun went down. They told tales of frontiersmen, tales of the Alamo, and tales of their forefathers. The rhythmic crickets, murmuring creek, a braying horse, all mixed harmoniously to create campfire music to accompany the stories. Flickering firelight jumped in emphasis with the words, the flames licking the boys' imaginations.

Camping, fishing and hunting trips were a frequent activity. To pass the time at the end of a day, they told tales of bears, ten foot grizzlies, and the brave ranchers who hunted and killed the vicious beasts, keeping women and children safe from harm. "And I've got the claw to prove it!"

They told tales of frontiersmen traipsing into unknown country—adventurers and their discoveries, the fossils they'd found still scattered about, mastodon and dinosaur bones crowding the window ledges and fireplace mantle at home. "Imagine huntin' this feller!"

Sometimes there were stories about Regulators and Moderators, stories of ballyhoos caused by violation of woman or principle, and tales of chivalry where the good guys were differentiated from the bad guys not by their title—sheriff or outlaw—but by their skill with a pistol, fairness at cards, and gallant treatment of women and children.

Granddad Carroll, pushing his worn and well-loved cowboy hat back on his head, still covered with thick grey hair, leaned forward with his elbows on his knees and told about cattle-drives that always started with, "You boys never worked a day in your life 'til you've driven a herd of longhorns in a rain storm in Texas."

The boys laughed at the tale of the lanky cowpoke driving cattle with Great-Uncle Columbus blazing a virgin trail north—naked except hat and spurs! He awed them with tales of cattle-rustlers and outlaws outwitted by trail-wizened cowboys.

Bob had listened to the stories longer than his brothers, and had become a seasoned teller of tales himself. He repeated ones he'd heard in his youth, embellishing here and there, to Art and Harold, cousins, friends, and the other boy scouts in his troop. He liked stories about Davy Crockett. When it came round to his side of the fire, he cleared his throat and spoke in a captivating voice, low and measured:

"In 1836, Davy Crockett arrived at the Alamo..."

Granddad Carroll had just returned from refilling his coffee and hearing this reference to the Alamo, muttered, "thirty-two men from Gonzales fought at the

Alamo. No other volunteers came, no one else." He took his seat on a camp stool and poked a stick into the embers sending up sparks.

Bob paused only slightly, knowing Granddad liked tales of "The King of the Wild Frontier." He continued, "Outnumbered twenty-to-one by Santa Anna's ruthless army they refused to surrender. Davy yelled: 'Give me victory or give me death!'

"I believe that was Patrick Henry." Charley said quietly, ignored.

"Crockett was all over the place in defense of the Alamo, it was like there were three of him. It was his bullet that grazed Santa Anna himself, knocked him right off his horse, sending him flying. And Crockett's crack shot killed Mexicans like sitting targets, like shooting bottles off the fence, ping, ping, ping!"

There was a chorus of, "Yeah!" "Yeah!" around the campfire.

"Now boys, we lost at the Alamo." Charley reminded everyone reluctantly.

They laughed and Bob went on with his story, "The only Texan that lived to tell about the battle was little Joe, a slave brought to the Alamo by Colonel Travis. Joe said the only one who surrendered, a cowardly fellow named Warner, was shot dead on the spot by Santa Ana's men."

Jewell, Bob, Myrtle, Harold, & Art

"Serves him right," someone spouted off, accompanied by sounds of general agreement from the campers.

Bob cleared his throat. "Davy and five or six of the bravest and finest shooters were all captured when they were trying to help others escape. They were dragged in front of Santa Anna himself. The general was angry, still bleeding from wounds caused by Crockett, and he couldn't believe his men had brought these gringos into his presence alive. He ordered them all bayoneted and then shot."

Logs settled and the fire hissed sending fireworks of sparks skyward. Someone sighed; Art made patterns in the ash with his toe. Harold yawned and stretched. Bob continued quietly, "And yet some still claim that Davy Crockett survived the Alamo. He was taken to the salt mines, deep in the Mexican interior, and made to work as a slave. He may survive there to this day." He finished his story with pizzazz.

"By gum, he'd be old," Granddad George muttered.

"Be always sure you are right, then go ahead," said Charley. His words mangled by the customary toothpick between his teeth that he gnawed in place of a cigarette.

"What?" Harold asked.

Charley removed the toothpick, "That's what Crockett said, 'Be always sure you are right, then go ahead.'"

Jewell Carroll, dressed to fish, Dayton, Wyoming

CHAPTER 4

REPROBATES & MALCONTENTS
You may all go to hell, I will go to Texas. ~ Davy Crockett

GRANDDAD GEORGE WAS A TRUE COWBOY, BORN AND BRED. HIS GRANDPARENTS HAD VENTURED WEST FROM the Carolinas into Texas, a wild frontier that didn't enter the union until 1845, and shortly thereafter tried to secede.

George's father, also named George after the first president of the United States, was a middle child born in Atlanta, Georgia, in 1832, to Jacob and Sallie Carroll. Jacob was selected to serve on a grand jury in 1831 and '33. His occupation is listed as "planter." He is recorded as a member of the Mount Tabor Baptist Church, one of the oldest churches in Georgia. And on April 4, 1836, Jacob Carroll, "a wealthy slaveholder in the Hightower district bought Town lots 11 and 36 paying $300." Lots 11 and 36 were human beings.[1]

We know little of Jacob and Sallie and slavery was never mentioned by later generations of the Carroll family. Some topics became taboo dinner table discussion long before Bob, Art and Harold sent green peas flying towards each other off the catapults of heirloom silverware. Some history is glorified and some is not acknowledged. The glorified part included remembrances of Davy Crockett.

Davy Crockett, "King of the Wild Frontier",[2] fought alongside General Andrew Jackson in the Indian Wars against the Creek Nation. Many of the Creek they were fighting were also of European descent due to decades of intermarriage, and it might have been difficult to identify an enemy by sight alone, as there wasn't much to distinguish one from the other. The

Creek, although backed by both Spain and England, were forced to give up 40,000 acres of land in 1814 after losing to Jackson's forces who won, in part, due to the help of the Cherokee.[3]

After the war, Crockett moved his family to Franklin County, Tennessee, and in 1821, he was elected to the Tennessee legislature. He served two terms, during which time he claimed to have killed approximately 105 bears.[*] In Congress, he made a name for himself by opposing President Andrew Jackson's Indian removal bill. Jackson pushed for the removal of all Indians west of the Mississippi but Crockett advocated for family, friends and constituents who'd traded and intermarried with the Cherokee, Chickasaw, Choctaw, Creek and Seminole, these considered to be the "five civilized tribes".

Crockett left Tennessee to lick his wounds after losing his bid for governor. (It would have been his third term.) Happily bound for Texas, he encouraged others to follow, and tipped his hat to the United States. Crockett was always one to go where the action was, and found his way to San Antonio de Bexar in early February of 1836. Mexican General Santa Anna arrived on February 20th.[4]

The Texas Crockett entered was owned by Mexico and developed by Empresarios.[†] The Empresarios received land-grants from the Mexican government which they sold at a profit. They distributed flyers by the thousands from New York to Ireland glorifying the virtues of Texas, often considerably embellished.

Empresario Martin De Leon, a Mexican native, colonized Mexican families along with a few Irish in the Guadalupe river valley beginning in 1824.

Green DeWitt, an American, led American families to settle along the Guadalupe in 1825, bordering De Leon's colony. But that border was unmarked. You can draw a line in the sand, but the Texas wind obliterates anything but the memory of it, and that can be disputed. Where their claims overlapped, the Mexican government favored De Leon.

Martin De Leon and his colonists were better prepared than DeWitt's people for the harsh environment and the never-ending ambushes by hostiles. Indians ruled Tejas, including the Apache, Comanche, Karankawan, Kiowa, Lipan, Tonkawan and others. The Tonkawas and the Karankawas each had permanent villages long established in the verdant

---

[*] Crockett published his autobiography in 1834 detailing his early life. Like other Scots-Irish, Crockett spent time in his youth driving cattle.

[†] An Empresario was someone who was compensated, usually in land, for bringing colonists or settlers into Spanish or Mexican owned Texas.

river valley the Americans were trying to settle.

In a letter to Mexican Official Saucedo, Stephen Austin lamented, "I conceive it to be my duty to inform your Lordship, that a few days since, a party of about sixty Comanche and Tahuacano Indians visited the settlement of the Empresario Green De Witt. In the town of Gonzales in the Guadalupe, they killed one man and wounded another destroyed all the furniture of Mr. Kerr's house, and stole all the horses they could find. The settlers retreated to this Colony, not knowing what to do; they had a certain quantity of corn sowed, and for this unfortunate occurrence, their Colony would be in a very thriving condition."[5]

In response to their difficulties, General Domingo de Ugartechea in San Antonio ordered one small cannon delivered to the De Witt's colonists in Gonzales for protection.

DeWitt eventually managed to settle 189 families, and their slaves, before his contract expired in 1831.[6] By then, the population of Texas had increased to over 40,000 Americans who'd negotiated for land legally or illegally.

The great migration into Texas led to an increase in slavery in the south at the same time that it was being eliminated in the north.* Mexico had recently passed a law that prohibited bringing slaves into Mexican-owned Texas but it was impossible to enforce—slaves were simply turned into "indentured servants" with some ink on paper.

Mexican General Teran described the Texas he saw in 1828: "The whole population here is a mixture of strange and incoherent parts without parallel in our federation; numerous tribes of Indians, now at peace, but armed and at any moment ready for war, whose steps toward civilization should be taken under the close supervision of a strong and intelligent government; colonists of another people, more aggressive and better informed than the Mexican inhabitants, but also more shrewd and unruly; among these foreigners are fugitives from justice, honest laborers, vagabonds and criminals, but honorable and dishonorable alike travel with their political constitution in their pockets, demanding the privileges, authority, and officers which such a constitution guarantees. Added to this motley mixture [are] the slaves beginning to learn the favorable intent of the Mexican law toward their unfortunate condition and held with an iron hand to keep them in a state of subjection."[7]

For their part, the colonists had mixed sentiments regarding their Mexican citizenship. Those in Gonzales and San Patricio wanted Texas to

---

* Northern states passed emancipation acts between 1780 and 1804.

remain part of Mexico. But there were plenty of "Muldoon Catholics" who were Mexican citizens in name only.[*] And Mexico's policy had changed toward the Texians. Santa Anna, as president, was clamping down on everyone. He had ordered all townships to dissolve their militias.

In response, DeWitt's colonists formed a group for defense and called it "the Safety and Correspondence Committee" which, in turn, led General Ugartechea to see his formerly-threatened Gonzales colonists as threatening. He rode with troops to take back the cannon he had given the colony for protection against Indians. And small though it was, the Gonzales cannon became the symbol for the Freedom of all Texas. Flying a flag that stated boldly: "Come and take it," DeWitt's colonists fought for and won their right to keep their cannon. The Battle of Gonzales is remembered as the first battle of the Texas Revolution although the Mexicans withdrew and there was no loss of life.

The Battle at the Alamo, however, was a fight to the death. Mexicans, Scots-Irish Texians, and others fought together to defend the small fortress near San Antonio, even though Santa Anna and his army were certain to win. The Texians were fighting against the restrictions of a dictator and despot. Santa Anna saw them as traitors, especially the Mexicans, and gave orders to take no prisoners.

Col. William Travis sent for backup as Davy Crocket, James Bowie, and others prepared to fight. Travis knew they were desperately outnumbered. Although Col. Fannin's troops were nearby at Goliad, no one was dispatched to the Alamo. But thirty-two men from Gonzales, who'd just fought to keep their cannon, rode hard and fast to San Antonio to join 150 or so others in defense of the Alamo.

The town of Gonzales filled with families awaiting word of the outcome of the battle. It was assumed that Santa Anna would attack them next. News finally came with a bedraggled and unlikely group, witnesses and survivors: Susanna Dickenson and her daughter, Ben and Joe who were slaves of Colonel Almonte and William Travis; and their guides Erastus (Deaf) Smith and Henry Wax Karnes. Together the six sad messengers delivered a letter from Santa Anna to Sam Houston in Gonzales. The next day, on March 13, 1836, Houston gave orders for the town to be burned. Thirty-two buildings erected in joy were now torched in sorrow. The grieving families headed east toward the Sabine River in what would be known as the Runaway Scrape.

General Sam Houston and a small group of battered Texas soldiers

---

[*] Being baptized by Father Muldoon of the Austin Colony didn't mean much except to the Mexican officials who required it.

continued eastward in retreat and took cover near the river in the trees. Santa Ana advanced toward them across a wide expanse of prairie. The Texians were seriously outnumbered but well positioned for battle and they were aided by the fighters that would later be called, "Texas Rangers." The Rangers covered the retreat of civilians, "harassed columns of Mexican troops, and provided valuable intelligence to the Texas Army," according to Mike Cox of the Texas Ranger Hall of Fame.

In addition to Texas Rangers, they also had passion on their side. Vengeance cries of, "Remember the Alamo!" could be heard on April 21, 1836 as Houston and 750 soldiers defeated the infamous Santa Anna and his 1500 troops at the Battle of San Jacinto.

Pumped with victory and riding high in his saddle, Jim Sylvester spotted the Mexican crouched in the brush, barefoot and wearing a corporal's uniform. He couldn't believe his eyes when he saw him flash the Masonic sign of distress. Because Sylvester was a Mason, too, Santa Anna was spared and taken prisoner.[8] He would take power again in Mexico.

FALL 1931

Bob, Art and Harold stand shoulder to shoulder, faces washed, hair combed, and all spruced up in their Sunday best. Standing next to Bob in the Sheridan Baptist Church is Myrtle, then Granddad George and Grandma Ella, Arthur stands next to fidgeting Harold on the other end. The organ belches "Shall We Gather at the River" slightly off key. The boys watch as Granddad George mimes the words from the hymnal, but no sound comes from his mouth. His eyes sparkle and he throws back his head as if to rattle the room with his resounding tenor. Bob covers his laugh while Art and Harold laugh out loud. Ella steps on George's foot, hard.

After church, they take advantage of the fine weather to picnic in their favorite spot on the river up above Dayton, near the old family ranch.

"Awe! I don't have my fishin' rod!" Harold complained, as they bumped along on the dirt road edging the river and came to a stop.

Myrtle laid out a fine spread, including corned beef sandwiches and freshly canned pickles. Art and Harold sat and munched loudly while Arthur made himself comfortable on a nearby rock and got out his pipe and tobacco. After the meal, Myrtle cleared food from the blanket, and then shook it to scatter left over bread crumbs to the birds, then folded it neatly and placed it in the picnic basket. Bob ambled over to his father, who sat in the shade of a yellow and orange leaved Aspen tree, blowing smoke rings and looking out over the valley below.

"How'd you use up all of the gas in the car yesterday, Bob?" Arthur asked when he was close enough to hear.

Bob blushed. He'd spent last evening here with a girl; the grass under the oak was still flattened from their bodies. They'd gone to a show and then for a drive here, a favorite place. "I took a drive last evening," he answered after a pause.

Arthur didn't reply.

"I'm sure you'll fill up the tank and clean 'er up when we get home," was his father's only comment.

"Yes, sir," Bob enjoyed the use of the car, but a tank of gas was a high price to pay for a little romance. Gas was an expensive commodity in 1931, ten cents per gallon! But he couldn't get around by horse like back in Granddad George's day. "I'll make sure the car looks spic and span, Dad."

When he thought about it Bob understood that life wouldn't be the same at all without Dad's beloved car. His whole existence would be different if they depended on horse and buggy and trains for transportation. Without automobiles the world was big and everything distant. The world suddenly became smaller and more accessible with autos. Bob wanted to go everywhere and see everything. He wanted to drive to the top of the mountain in ten minutes with his arm around his girl!

In contrast, Granddad George told about his father crossing into Texas. No short buggy ride. Bob couldn't imagine riding horseback across Wyoming much less riding from Georgia to Texas by way of Arkansas. What an adventure that must have been!

Jacob Carroll, George's grandfather, rode ahead with his older sons, Dennis and Louis on the lookout for hostiles and outlaws. Ten year old George Washington tended the horses. Jacob's wife Sally, their daughters and younger children, with the help of slaves, walked along beside their belongings and kept the wagon wheels out of the mud and on track. Few traveled by horse or rode in the wagons.

They traveled in a big caravan with Jacob's brother Dennis, and his family, and the Cockburn family, at a rate of about three miles per hour, if they were lucky, totaling perhaps fifteen miles per day. Four to six steer pulled wagons loaded with all of their possessions, including the parts necessary to construct a cotton gin and an empty wagon for security.

Young George Sr. saw many amazing things during this journey that stayed with him for the rest of his life and became the nuggets of long-told tales. The most amazing sight was the silver dust cloud on the horizon that kept moving faster and growing larger. "Look there, Son! Mustangs!" Jacob tipped his hat toward the shimmering apparition. When the herd finally became visible, George was awe struck. There were so many! Their sinewy muscles slick with sweat, they were of different colors and patterns but they seemed to move as one and with an unfathomable determination. Charging across the prairie like they had somewhere to go! Although they were gone in a moment from sight, they left George forever changed. He was going to like this Texas.

Abner McElroy & Ellenor Kuykendal McElroy, about 1835

CHAPTER 5

THE INCURSION OF THE CLANS
Migration to the New World and along the American frontier was a 'movement of clans' who, although immigrating at different times, tended to settle together in the American backcountry. ~ Joshua Lee McKaughan

Most of the Carroll men were Freemasons, like many other Scots-Irish, and joined with other Freemasons wherever they lived. They created communities through friendship and intermarriage. In America, the Scots-Irish often intermarried with the German and Dutch. Follow the family names — Carroll, Cockburn, Shipman, Van Kuykendall, Van Zandt, Westfall — to find the patterns of consanguineous marriage (where the couple getting married have a common ancestor), that tied this group of people together for generations, building layers of relationships across time and space.

Like other Scots-Irish, Jacob and Dennis Carroll Jr. moved from the Carolinas into Georgia and Tennessee, entering the Blue Ridge and Appalachian mountains with their families. They built spacious log cabins with giant fireplaces where they burned the forests they'd clear-cut. When they moved, they left behind their cast-off, handmade household and personal items, like carved wood combs and pipes. They passed down many positive attributes to their descendants and their adopted country: love of freedom and self-sufficiency and a strong sense of fairness and equality. They also passed down a wasteful land management policy— use it until it's gone and then move on. The practice of slash and burn agriculture required the Scots-Irish to move frequently, and they migrated across North America east to west.

In Tennessee, Davy Crockett was friends and neighbors with revolutionary war veteran, Jacob Van Zandt. Thirty-five years older than Crockett, Jacob Van Zandt was a respected citizen of Tennessee. He'd served as Captain in the North Carolina Militia, in the Revolutionary Battle of Cowpens (Cowpens: a place where cattle were kept) at the end of the Southern Campaign in 1771. This was considered a turning point of the war compelling the English to surrender at Yorktown.[9]

The English should have known they were fighting a losing battle. Many who fought in the Southern Campaign for the United States were Scots-Irish. The English had already lost to this enemy and had a name for them: "Reevers" (or "robbers") named for their successful combat in the Border Wars between England and Scotland. These border warriors, or Reevers, left Scotland and went to Ireland. As Ulster (Irish) Scots, they continued to fight the English for religious freedom and affordable housing. Many moved to America where the term 'Scots-Irish' was created to distinguish them from their Catholic neighbors. During the American Revolution, the English met the Reever's children and grandchildren, toughened by the Indian wars, and just as strong-willed as their predecessors.

Davy Crockett was also acquainted with well-known tavern owner and veteran, Abraham Kuykendal, Sr. Like Van Zandt, Kuykendal was a captain in the North Carolina Militia from 1770 to '83. In 1779, he served as justice of the peace in North Carolina.

Kuykendal owned a popular tavern near Flat Rock and a lot of land, some of it given to him because of his Revolutionary War service. He had eleven children with his wife, Elizabeth, before she died in childbirth. Then he married the beautiful and much younger Bathsheba, and had four more children. After he'd died, tales were told of Abraham's ghost.*

Abraham Kuykendall's son, Abraham, married Jacob Van Zandt's daughter, Elizabeth, in 1790.[10] Abraham's and Elizabeth's grandchild, Nancy Abigail "Abby" McElroy, would marry George Washington Carroll and weave all three families together. Bob Carroll was close to his Great-Grandma Abby until she died when he was ten.

George and Abby's oldest son was Granddad George—George Granville Carroll. His middle name, Granville, doesn't commemorate a grandparent, as is usual, but instead it commemorates the English Earl of Granville, a Carolina landholder, and marks the origins of the Carroll migration into the Carolinas. †

---

* The Tale of Abraham's Ghost (extrapolated from many versions on Ancestry.com and other sites) is reproduced in the Appendix A.

† See Appendix B for more about the Jacob Carroll migration.

Unknown Girl, Fairbury, Illinois

CHAPTER 6

TEA-TIME ON THE FRONTIER
Possessions, especially among the more ambitious, equal wealth and status.
~ Joshua Lee McKaughan

Leaving Granville land behind them, its fertile soil depleted, Jacob and Dennis Carroll Jr. sold their property and moved into Cherokee territory—Forsyth County, Georgia—in 1828.

They migrated with their families. Dennis Jr. had married Nancy Waggoner in 1814, and they produced a brood of children, of note: Celia, Sarah, Jacob, Drury, and Martha.[11] Jacob married Sallie sometime before 1822 and their first son, Dennis, named after Jacob's father, was born in 1822. Second son, Lewis, was born in 1825; Mary in 1827; George Washington (Granddad George's father) in 1832; Elizabeth (Lizzy) in 1838; Jacob Jr. in 1839; little Sallie (named after her mother) in 1840; and finally Columbus—named after the explorer and just as adventurous—was born in 1843.[12]

In 1831 settlers were required to sign an Oath of Allegiance to the State of Georgia that allowed the signer to continue residence within the Cherokee Nation. Jacob's and Dennis Carroll, Jr.'s signatures do not appear; however, they both served on a grand jury in Forsyth Superior Court in the early 1830's. Both Jacob and Dennis were members of the Mt. Tabor Baptist Church,[13] and in August 1833, the Church was moved to the home of Dennis Carroll Jr. and his wife Nancy. Charter members included both Dennis' and Jacob's families and their slaves, including a man named York.* For unknown reasons, Jacob was excommunicated from the Mount Tabor Baptist Church in 1840.

Jacob and Dennis Jr. migrated with their families into Cherokee country between 1831 and 1838. At this time, the Cherokee people were being forced from their homes by a series of laws that drove them west of the Mississippi. We call this the Trail of Tears today because many people died. Migrating white settlers were entitled to take over or "purchase" the land where Cherokee and Creek Indians had lived and hunted for generations.

"Some Indians had already emigrated to the west and some stayed as long as they could or until they were forced to leave. Those who did not emigrate were marched to Fort Campbell (formerly Camp Gilmer), put into the compound, then marched to Ross' landing (Chattanooga, Tenn.) for transporting to Oklahoma and Arkansas."[14]

Jacob and Dennis Carroll Jr. appear to have been personally responsible for driving one family from their home. According to the Indian Valuations in Forsythe County from September 28, 1836, Alfred H. Hudson, a white man, and Susanah (Buffington) Hudson, his wife, a quarter-blood, lost the spread they'd worked to improve. An afidavit, signed by evaluator, L. J. Hudson, states: "Alfred H. Hudson was dispossessed of the whole of his improvement on red bank creek in said county by Dennis Carroll in the early part of 1835."[15]

The Hudson property that Dennis aquired was located on Red Bank River near his own property. Alfred and Susannah Hudson had owned and improved several plots of land which included a house, farm and orchards. Jacob took over this property and became neighbors with his brother Dennis.

During the time that Jacob and Sarah were making themselves comfortable on the Hudson property, Davy Crockett left for Texas and the Alamo.

Jacob and Dennis Carroll Jr., and their families resided in Georgia only a few more years and then they followed Crocket's trail, traveling to Arkansas in a big caravan of interrelated families of Carrolls, Cockburns, and Shipmans.

The Carroll family migration essentially followed the same route as Davy Crockett's—"down the Mississippi River to the Arkansas and then up that river to Little Rock; overland to Fulton, Arkansas, and up the Red River along the northern boundary of Texas; across the Red River, through

---

* His slave name may come from the place where he was born or purchased.

Clarksville, to Nacogdoches and San Augustine; and on to San Antonio."[16] The close connection with Davy Crockett passed down into Carroll family names to keep the memory of their relationship to Crockett alive.[17]

The Arkansas they entered was hardly a wilderness. Thomas Nuttall travelled the Arkansas River and gave a detailed account of the Western Cherokee in *A Journal of Travels into the Arkansas Territory* During the Year 1819. He describes a lush valley: "Both banks of the river as we proceeded were lined with the houses and fences of the Cherokee, and although their dress was a mixture of indigenous and European taste, yet in their homes, which were decently furnished, and in their farms, which were well fenced and stocked, we perceive a happy approach toward civilization. Their numerous families, also, well fed and clothed, argue a propitious progress in their population. Their superior industry, either as hunters or farmers, proves the value of property among them, and they are no longer strangers to avarice and the distinctions created by wealth. Some of them are possessed of property to the amount of many thousands of dollars, have houses handsomely and conveniently furnished, and their tables are spread with our dainties and luxuries."[18]

Like the Cherokee, the Carrolls and other Scots-Irish strove to keep their network of family and allies together in spite of the chasm created by the distinction between the "haves" and the "have nots." Although they were tied together by blood and a similarity of circumstance, they were divided by the "baubles of Britain,"[19] the imported goods from the United Kingdom that created status.

The "haves" enjoyed tea-time with quality imported English tea sets. The "have nots" were identified by their cracked teacups and missing sugar bowls, or no tea time at all! Those who indulged in tea time on the frontier, and had the paraphernalia to do so, might flout it before their poorer cousins, and were seen as putting on airs, creating the same kind of class-society many had come to America to escape.[20]

It is possible that class distinction contributed to Jacob and Dennis Jr. going their separate ways once they reached Arkansas. The Jacob Carroll family continued on to Texas. In Texas, they would reconnect with friends and relatives: the Kuykendalls, Van Zandts, and Shipmans.

The Van Zandt and Kuykendall families were friends and neighbors of the Shipman family. We can follow Daniel Shipman's history, recounted in 1879, to discover more about the Carroll migration. Daniel Shipman was one of the "Old Three Hundred," so called because they were among the first settlers in Texas. He chronicled his birth in North Carolina at the turn of the 19th century, the Scots-Irish route he and his family traveled, and his "making crops" in Tennessee. He went to Texas by way of Arkansas and explored the Brazos River region. Daniel Shipman joined Austin's colony in the fall of 1823, then joined Martin De Leon's colony, but later returned to the Brazos and married Margaretta Kelly in 1828. He fought the Karankawa Indians[21] during the summer of 1825 and, after he was no longer able to do so, he recounted his exploits in the book *Frontier Life: 56 Years in Texas*.[22]

Above Story, Wyoming in 1927. Back: Jim Carroll, Art Jr., Arthur. Front: Harold, Myrtle, & Bob.

CHAPTER 7

THE REPUBLIC OF TEXAS

They were men who could not be stampeded. ~ Col. Homer Garrison, Jr., Texas Department of Public Safety, speaking of Texas Rangers

TEXAS FLEW HER FLAG AS A FREE REPUBLIC FOR TEN YEARS. THE FIRST PRESIDENT OF TEXAS, MIRABEAU LAMAR, OFFERED 640 acres to those willing to become Texas citizens in 1838. Jacob and Sallie Carroll took him up on the offer. Lamar also put a bounty on every Indian in the Republic and declared that free blacks were not allowed to reside within the territory, strengthening the power of slaveholders.

Jacob and Sallie built their plantation within the original boundaries of the DeWitt Colony, just thirteen miles south of the town of Gonzales. This was the heart of the beautiful Guadalupe River Valley, near where the Guadalupe and the San Marcos Rivers join, and the lush banks were lined with stately, old walnut trees filled with wild game and abandoned Spanish cattle.* [23]

The main plantation house was laid out in the traditional style, two stories with the second supported by spaced columns creating a porch in front. There were ten or twelve rooms, circular gardens adorned the front lawn, slave quarters and outbuildings in back. There were orchards and

---

* When the Carrolls left the valley after the civil war the appearance and character of the place had completely changed. "A firm from Ohio purchased 1,000,000 feet of walnut timber on the Guadalupe River between Gonzalez and Belmont in 1882...The demand for walnut timber was so great that the trees were completely cleared from the Gonzalez County river bottoms within a few years." The History of Gonzales County Texas.

berry patches in addition to the cotton and cattle Jacob tended while Sallie tended children. When sitting on the front porch of the house, one could view the country over the entire valley of the Guadalupe. Jacob and Sallie Carroll are buried here in what was the Carroll family cemetery in current Glaze city, as are their daughter and son-in-law, Sallie and James Turk.[24]

Their neighbors were German, Irish, Mexican—a whirl of peoples with varied backgrounds and cultures who responded differently to life's ups and downs. In Orange County, the settlers were mostly "old emigrants from Southern Louisiana and Mississippi and more disposed to gayety and cheer than the Texas planters [of the Brazos region]."[25] Some danced and sang when the harsh sun went down, others stood straight backed, stoic in their gloomy perseverance. Regardless of their differences they all shared in the inescapable isolation, as constant as the wind.

Jacob Carroll and his older sons, Dennis, Lewis, and George Sr. were frequently on the trail. They sold their cattle and crops and purchased slaves and supplies in New Orleans.

The "Isle d'Orleans," as the French referred to it, was both a physical and cultural island. Founded in 1718 by the French and later owned by the Spanish, it was finally sold to the Americans with the Louisiana Purchase in 1803. A cosmopolitan city based upon European design, New Orleans was populated by Choctaw, French, Spanish, Creole, English, German, Irish, free Africans, and others. There was a great distinction based on background and experience between free Africans and slaves whose skin might be of the same hue. Free Africans were typically Catholic, spoke French and were often respected citizens and property owners, sometimes owning slaves themselves. Free citizens of color also served in the American military in the war of 1812 and later in the Confederate forces.[26]

Jacob and Sallie Carroll became wealthy growing cotton in Texas. Jacob sent money back to his brother, Dennis Carroll Jr., for the construction of a library in Montgomery County, Arkansas that bears the name Carroll for them.[27]

They did not live peacefully in Texas. Lonely spells were broken up by encounters with outlaws, hostile Indians, and border fights with Mexicans. Texas claimed the border with Mexico was the Rio Grande. Mexico insisted the border between them was the Nueces River. Santa Anna threatened that U.S. annexation would be seen as an act of war. U.S. President Polk ignored this warning and sent American troops to the Texas border in 1846. The two year Mexican-American War resulted in Texas becoming part of the U.S., as well as New Mexico, Arizona, Nevada, California, Utah, and parts of Colorado, Wyoming, Kansas and Oklahoma.

Following the war with Mexico, stories were told of Los San Patricios. These were U.S. soldiers, mostly Irish immigrants and Catholic, who were offended by the inhumane behaviors of some wearing U.S. uniforms during the two-year war. When they saw Catholic churches being pillaged and burned they "turned coat" and fought with the Mexicans. The U.S. court-martialed almost one hundred San Patricios for deserting after Mexico's surrender. Fifty men were hanged. In Mexico, they are still revered as heroes.

Then the floodgates opened and Americans started streaming across the border. "The Texas to which these migrants came was a frontier state in the classic sense. That is, it had a line of settlement advancing westward as pioneers populated and cultivated new land."[28]

The Army built Forts, mostly completed in 1852, to protect the settlers, but Indian warfare continued even after the Civil War. "The Comanche and the lack of water and wood on the western plains both hampering its advance, the Texas frontier did not move during the 1850s beyond the seven forts completed at the onset of the decade. Areas immediately to the east of the military posts continued to fill, but the rush westward slowed. In 1860, the line of settlement ran irregularly from north to south through Clay, Young, Erath, Brown, Llano, Kerr, and Uvalde counties."[29]

The Carroll family lived continually on the edge, along the line of settlement running north to south. The edge so sought after, fought for, conquered and controlled may have existed not only as an outer landscape but also an inner landscape affecting the boys' vision and view of the world as they grew. Bob, Art, and Harold learned to see life and their place in it through a prism, the refracting light emphasizing an atmosphere, an attitude, and a way of life.

SPRING 1932

Snow blanketed Sheridan in quiet isolation. It covered the roadways and rooftops and treetops, even the magnificent climbing tree that shaded the Carroll's tiny well-maintained home across from the grade school. Everything was covered with feet of fluffy white powder. Borne by the wicked winter wind, it had curled around corners to layer unevenly on window ledges, lapped over porch screens to create rippling patterns, and drifted into mounds that covered bushes and bicycles, burying whole cars under its weight. It covered Myrtle's flowerbeds leaving only some tall stalks visible, outlined in glistening ice.

Sledding behind cars was prohibited in Sheridan. And so Bob, Art, Harold and others got out their skis. They gathered at the top of 14th Street hill, put their skis on and waited, throwing snowballs and shoving each other around to stay warm. They were waiting for the first likely unsuspecting bumper to come their way.

Art pulled his cap down over his ears, rubbed gloved hands together, and grabbed onto the silver fender of a blue Oldsmobile sliding off on wobbly but confident skis. He guided his way nimbly along the edge of the road, slipping back behind the car to avoid heaps of snow, trash cans, and occasional pedestrians. The hill was considerably steeper than he'd expected. His scarf flapped in the wind like a kite tail as he shifted his weight expertly beside the Olds. Art was fourteen and fast. But gaining speed wasn't the hard part, slowing down was. At a T-intersection, he lost the car when it turned the sharp corner to the left and Art found himself continuing forward, flying.

He winced a bit now as he stretched, still sore, but he was the first one up as he usually was on cold mornings, up before anyone scratched a match to the cast iron to light the wood stove, when fingers of ice still tickled at the spine, encouraging a shiver. This morning was no exception. He nursed the cut that was sure to turn into a scar and remembered the impossibility of being airborne and the pain of landing as he scuttled into his jeans.

There was no point lying around in bed. It was a special day anyway. The question was: how best to wake up his lazy brothers? He thought of pounding out a few notes on the piano; Myrtle was so proud of his progress. But the piano sat in the living room, closer to Mom's and Dad's room.

Then it hit him. Art cleared his throat loudly and in booming tenor recited his favorite cold weather poem to Harold's great distress, and soon, to Bob's. Although he was in a separate bedroom, only a thin wall divided them.

"There are strange things done in the midnight sun by the men who moil for gold..."

All three boys possessed lovely voices, but at certain times even the best tenor is unappreciated.

"Be quiet!" the wall bellowed.

Art continued on, redoubling his efforts to recite great literature with flair on this cold, damp dawn while adding a second pair of socks and lacing up his boots.

> "The Arctic trails have their secret tales
> That would make your blood run cold;
> The Northern Lights have seen queer sights;
> But the queerest they ever did see
> Was that night on the marge of Lake Lebarge
> I cremated Sam McGee."

Art put on a clean flannel shirt while ducking pillows and other objects coming from Harold's side of the room. He was just warming up.

> "Now Sam McGee was from Tennessee, where the cotton blooms and blows.
> Why he left his home in the South to roam 'round the Pole, God only knows.
> He was always cold, but the land of gold seemed to hold him like a spell;
> Though he'd often say in his homely way that 'he'd sooner live in hell.'

> "On a Christmas Day we were mushing our way over the Dawson trail.
> Talk of your cold! Through the parka's fold it stabbed like a driven nail.
> If our eyes we'd close, then the lashes froze till sometimes we couldn't see;
> It wasn't much fun, but the only one to whimper was Sam McGee."

"You'll be whimpering when I get a hold of you!" the wall threatened. Harold retrieved a pillow, and now back in bed, covered his head with it and whimpered loudly.

> "And that very night, as we lay packed tight in our robes beneath the snow,
> And the dogs were fed, and the stars o'erhead were dancing heel and toe,
> He turned to me, and 'Cap,' says he, 'I'll cash in this trip, I guess;
> And if I do, I'm asking that you won't refuse my last request.'

> "Well, he seemed so low that I couldn't say no;
> then he says with a sort of moan:
> "It's the cursed cold, and it's got right hold
> till I'm chilled clean through to the bone.
> Yet 'taint being dead—it's my awful dread of the icy grave that pains;
> So I want you to swear that, foul or fair, you'll cremate my last remains.

> "Now a promise made is a debt unpaid, and the trail has its own stern code."

Art paused, a shuffling sound could be heard behind the wall, followed by thumps rapidly approaching. He ducked behind Harold's bed just as Bob burst through the door and kept coming. Up and over Harold's bed, ignoring the "Ow!" coming from the pillow, Bob leaped onto trapped Art with a thud. A scuffle ensued. It was short lived as Art was quickly hog-tied and begging uncle.

"What? What? I can't hear you!" Bob insisted until Dad appeared in the doorway. Arthur Carroll stood in the dimly lit hall with a scowl on his face. He said nothing for a moment while Bob released Art, and they both got to their feet, along with Harold who had been watching and shouting encouragements alongside the pair.

"I see you boys are getting ready to go. Looks like you're giving your brother something to remember on those lonely days in camp this summer. We'll be leaving in thirty minutes." Arthur spoke quietly, a slight twinkle in his eye, then continued down the hallway. Bob left the room glaring and pointing at Art in warning. Harold hopped into his pants. Art brazenly continued his poem, in a slightly lower voice:

> "Some planks I tore from the cabin floor, and I lit the boiler fire;
> Some coal I found that was lying around, and I heaped the fuel higher;
> The flames just soared and the furnace roared—such a blaze you seldom see;
> Then I burrowed a hole in the glowing coal, and I stuffed in Sam McGee.
>
> "Then I made a hike, for I didn't like to hear him sizzle so;
> And the heavens scowled, and the huskies howled,
> and the wind began to blow.
> It was icy cold, but the hot sweat rolled down my cheeks,
> and I don't know why;
> And the greasy smoke in an inky cloak went streaking down the sky.
>
> "I do not know how long in the snow I wrestled with grisly fear;
> But the stars came out and they danced about ere again I ventured near;
> I was sick with dread, but I bravely said: "I'll just take a peep inside.
> I guess he's cooked, and it's time I looked;" . . . then the door I opened wide.
>
> "And there sat Sam, looking cool and calm, in the heart of the furnace roar;
> And he wore a smile you could see a mile, and he said: "Please close that door.
> It's fine in here, but I greatly fear you'll let in the cold and storm—
> Since I left Plumtree, down in Tennessee, it's the first time I've been warm."[30]

"You're going to feel warm!" the wall threatened again, "I'll sick the Yocum gang on you!"

The boys crowded their way eventually into the car, and the Carroll family traveled on a crisp clear morning down a newly snow-plowed road, their destination Cheyenne. Bob was on his way to enlist in the Civilians' Military Training Corps or CMTC.

The CMTC was a university training program that had been developed before WWI and was still used to prepare young men for military service. Bob and other volunteers would be given four weeks of military training each summer, starting this summer. Those who finished four years and home study courses were then eligible for commissions in the Officers' Reserve Corps.[31]

Bob dreamed of becoming a Commissioned Officer; it would be his leap into the future. Even though poverty's plague was all around him, and even though a university education was out of reach financially for his family right now, his skills and talents in the military would allow him to step up, step above, and seize his future with both hands. Myrtle beamed at the brave, handsome, and skilled young man her oldest son had become.

Although thoughts of the future were on Bob's mind as they drove, along the way the boys told stories, gorier and gorier stories about the past, tales of the Wild Bunch, tales of Al Capone, and tales of the Yocum Gang until Myrtle had to insist that they stop.

Art Carroll

George Carroll, Sheridan, date unknown

CHAPTER 8

REGULATORS, MODERATORS, OUTLAWS AND OTHER HEROES

The Regulators went to such extremes in their attempts to break up the outlaws that a group of countervigilantes came into existence to "moderate" the Regulators... both sides drew in friends, relatives, and sympathizers from many miles away. ~ The Handbook of Texas Online [32]

GRANDDAD GEORGE WAS RAISED IN ACCORDANCE WITH THE SAME STRICT BAPTIST PRINCIPLES THAT JACOB AND Sallie had impressed upon his father, George Washington Carroll. "Spare the rod, spoil the child" was an adage passed down through the generations. They'd required of their children the same obedience they'd required of their slaves. And a religious argument might even be heard in the well-ordered Carroll household in support of slavery.[33]

George and his brothers and sisters received no formal education; Sallie provided the education for her children, and was likely more demanding than any school marm. In addition to the Bible, the children read poetry and the classics. Jacob took part in the education of his boys in geography and history, and of course the more practical knowledge of raising crops and animal husbandry. Male slaves were also educated in this area, and often worked side by side with Jacob and his sons Dennis, Louis, George, and Jake Jr., branding cattle, shoeing horses and birthing both. The girls, Mary, Elizabeth (Lizzy), and Sarah (Sallie), and the household slaves were all taught reading and writing as a necessity, like knowing how to swim. Only the youngest son, Columbus, managed somehow to avoid learning how to swim—which he would regret later in life.

Some slaves were very skilled horse and cattle hands, and people treated them with respect. "Black slaves had tended cattle, usually on foot, in the colonial Old South… Because of their value as property, slaves were sometimes treated differently. Abel 'Shanghai' Pierce recalled breaking horses in Texas in 1853. Several slaves assisted. A superior ordered Pierce to break the most dangerous mounts. The slave owner did not want to risk injuring slaves, because 'those Negroes are worth a thousand dollars apiece.'"[34]

In spite of what some might have seen as relatively good living conditions for the slaves owned by the Carrolls, these were people bought and sold at market. Many had been previously separated from family and lamented the loss of their loved ones, mothers missing their young children, sons missing their fathers, wives missing their husbands—never to be seen again. Some spoke little English having recently arrived on ships to New Orleans from Africa where they were treated like animals, or cargo, stuffed like so many sacks of wool crushed together in the hold, and reeking of their own and their fellow prisoner's bodily fluids.

Often strong in body and spirit, they found ways to deal with their misery. Music was one way to sooth the soul. Songs and stories from their home villages could be heard at night coming from the slave quarters.

But the Carroll plantation was now their home and most worked, like most family members worked, to secure and protect it. Living near the Mexican border meant folks had to be kept safe from marauders; living on the frontier meant folks had to be protected from Comanche raiders. Law abiding folks were advised to be vigilant; outlaws were everywhere. Neighbors pooled their resources, lent a helping hand, banded together. Stories of the Yocum gang were told to children to make them behave.

Thomas D. Yocum had found infamy robbing travelers with his gang along the Natchez Trace in Mississippi about 1815. Encouraged by the local sheriff, he moved to Louisiana and continued the shenanigans, but by 1830, "T.D. Yocum had been driven from the Atascosita District, and his cabin had been burned. The Liberty Alcade reported to Stephen F. Austin that Yocum had murdered a slave father and stolen his family." The Yocum's fled to Jefferson County, Texas.

Tommy Yocum got on pretty well in Texas. "In 1838, he was summoned to jury duty, and his inn was designated as the polling place for the voters of Pine Island precinct. By 1839, he had acquired a herd of thirty horses and 500 cattle, and by 1840, was postmaster of the Pine Island post office."

But old habits die hard. In 1841 Tommy and the Yocum gang robbed and murdered twenty men, "mostly cattlemen, who were returning from New Orleans with bulging money belts." Yocum was discovered as the scoundrel he was and asked to leave the county. "He refused, however,

REGULATORS, MODERATORS, OUTLAWS AND OTHER HEROES | 47

and one hundred and fifty vigilantes, mostly from Liberty County, converged on Pine Island Bayou in September, 1841, burned Yocum's Inn, and drove his wife, children, and slaves from Jefferson County. A posse trailed and captured Yocum near the San Jacinto River in Montgomery County. Perhaps aware of the Yocums' previous fortunes with juries (T. D. Yocum's father bought acquittal from murder charges on seven occasions with perjured witnesses) the Regulators shot Thomas Yocum five times and then disbanded.

"Yocum's son who had fled with his father, returned to Beaumont to visit his wife. Chris, described as the 'best of the Yocum's' and perhaps not implicated in the murder ring, had served honorably in the Texas army for one year. Sheriff Robert West, aware that thirst for retribution still lingered at Beaumont, arrested the youth and locked him in the county's log house jail on January 15, 1842. The following morning, the lawman found young Yocum hanged to a nearby oak tree with a ten-penny nail driven into his skull."[35]

These were Regulators at work. And the Regulators were busy that year of 1842, the year they were given a name.

Later that same year and not far away…

> Charles W. Jackson stormed into Judge John Hansford's courtroom, rifle in hand. Marshall, Texas had never seen such arrogance and self-assuredness. Tagging along with Jackson was the sheriff who supposedly had him under arrest. July 14, 1842 is a day to remember because Jackson brought his own posse with him. Behind Jackson and the lawman, 150 men, armed to the teeth, filed into the courtroom. They lined the walls, filled the corners, aisles and all rows, to send a message to the judge.
>
> "Why," the judge asked the sheriff, "is Mr. Jackson being allowed to carry a rifle into my courtroom? You," his honor addressed the lawman, "are going to be heavily fined for this."
>
> Jackson strode up to the judge, lay the weapon on his desk, pushed a chair up to it, and took off his jacket and shoes, jumping into the chair.
>
> "Let the trial begin!" Jackson demanded, glaring down at the judge. Jackson knew exactly what he was doing. Hansford mumbled, "Court is adjourned till 9 tomorrow." Hansford went to a friend's home and curled up around a bottle of whiskey, since Marshall, Texas didn't sell alcohol then.
>
> Jackson was acquitted and organized the Regulators, a vigilante committee to suppress crime. Their idea of stopping crime included beatings, arson and murder. To stamp out this lawlessness, the Moderators formed. Some of the skirmishes included hundreds of participants. When Jackson lost his life, the war really hit its fever pitch. Hired guns were brought in from Austin to exterminate 17 Shelby County citizens. People died in Marshall, Texas inside City Hall. Combatants met their maker in hotels, city streets, dance halls and

even churches. Communities split in half over the Regulator Moderator issue. Those with the means built their own forts for protection.[36]

The Regulator-Moderator War of East Texas lasted from 1839 to 1844. Some said land disputes were the root cause. "The land disputes fueling the fight were brought on by the greedy that counterfeited land grant deeds for property they wanted that belonged to someone else. Deeds were traded in saloons for as few as 12 drinks. These extortionists included ranchers, bankers, judges, and officers of the law, the military and preachers. The less respectable riff-raff stole anything not nailed down." Houston finally sent in 500 militiamen to end the fighting.[37]

Jacob wondered if the feud was truly ended or just put off for another day. He could understand the fierce principles of the Regulators and the balancing sentiments of the Moderators. Like other Scots-Irish frontiersmen he was distrustful of central authority, relied on himself and "strong local leaders," and preferred to "take the law into [his] own hands rather than waiting for a solution to come down from above."[38] He also saw himself as a civilized and god-fearing man, and was glad to be removed from the fray and living out on the edge of the frontier. Elder sons, Dennis and Louis, who served when called as Texas Rangers, (although the term "Texas Ranger" was not used until much later) provided protection to the family. Dennis was an expert marksman and carried his Colt naturally at his side. He could shoot shotgun in one hand, Colt in the other.

It was said that a Texas Ranger could "ride like a Mexican, trail like an Indian, shoot like a Tennessean, and fight like the devil." Albert Bigelow Paine in his classic book, *Captain Bill McDonald: Texas Ranger* tells of the ranger's assignment to prevent a scheduled prizefight. At the train station the city's anxious mayor asked him: "Where are the others?" McDonald replied, "Hell! Ain't I enough? There's only one prize-fight!" Rangers took up arms only when needed to chase down outlaws or Comanche and defend their neighbors. Rangers also came to the aid of Sam Houston and brought an end to the regulator-moderator feuding.

SUMMER 1932

Myrtle had left the photo album out and opened at the kitchen table. A warm evening breeze blew in the sounds and smells of the railroad yard. The pungent odors of burning coal billowed out in clouds from the engine car, then dissipated slowly over the low row of houses. Mixing with the scent of pot roast and brisket, it wafted out to the workers and called them home for the night. Men's voices drifted in, their words unintelligible, as they tried to shout over the clanks and groans of metal connecting and disconnecting, steel rolling against steel.

She was preparing a meatloaf and listening more to the radio than the sounds from the rail-yard, or the conversation between Bob, Art, and Harold in the same room. The radio announcer occasionally referred to the troubled times they all faced. Money was certainly not abundant for the Carrolls in 1932, Myrtle agreed. Life was tough all over, including over-seas, the announcer proclaimed. Poverty was weakening a once democratic Germany. Germans had come to depend upon the U.S. economy that was now sinking. Myrtle added stale bread crumbs to her meatloaf to make it go further.

"Dang!" Art swore. "How'd you get him to make that pose?" he teased Harold, pointing at a photograph of Bob.

"I dunno," Harold replied.

"I guess you're just the photographer of the family," Bob answered.

Art snickered, "Well he has to take the pictures, look at the ones of him—with that goofy smile!"

"What do you mean?" Harold was not quite sure if he was being ribbed or not.

"I just mean you need to be out of as many pictures as you can. That's why you take them so well; you need to be on the other end of the camera!" Art ducked a punch from his little brother.

"Look at you talking, with your ears!" Bob defended Harold. "And look at that handsome devil." Bob pointed to a picture of himself.

A smiling Bob was pictured in uniform with his dog. Next to the photo there was a newspaper clipping: "Bob Carroll achieves the eagle rank in scouting."

Eagle Badge Is Given Young Bob Carroll Tuesday

An interesting event in the life of a boy scout was enacted Tuesday noon when Bob Carroll, a nephew of R.E. Carroll, was presented his Eagle badge before members of the Lions club at their weekly luncheon gathering.

Lion Roy Coonfield, local Scout executive, made the presentation after outlining a few of the accomplishments a Boy Scout must attain in order to secure the Eagle rank, which is the highest rating in Scouting.

Lion R.E. Carroll is Scoutmaster of Troop No. 7 of which young Bob is a member, and was honored during the presentation.

Bob still remembered Roy's words and mimicked them now for his brothers. "Robert, your conduct along the trail has been excellent. As an Eagle Scout, you become an example in your community. The torch you carry is not only yours but is ours also. May the oath you have taken remain graven on your heart forever." He saluted the refrigerator, and Art and Harold followed suit. Then they all broke down laughing until Myrtle shushed them. "I can't hear the radio," she complained.

"Come on fellas," Bob said. "Quiet down. Let mom listen to the radio." Art and Harold shoved him and shushed each other, then turned the page.

Robert E. was more than Bob's uncle and Scoutmaster; he'd acted as his mentor providing a reading list and a willing ear. But recently he admitted he had nothing more to teach. Bob now possessed an extensive library and the intellect to match. He was thought of as the brain of the family, expected to achieve. He'd reached the highest rank in scouting, had high marks in school graduating early and with honors— what more was there to accomplish here in Sheridan? His striving spirit and restless nature demanded more than a small town could offer. He was antsy to go to Ft. Logan, in a hurry to get on with his life.

"You look like a Ranger in that one, Bob," Art insisted. They all looked closely at the photo taken on a narrow bridge, Bob in a cocky pose and wearing square toed riding boots and a cowboy hat slightly askew.

Robert C. Carroll, boy scout

Bob smiled. He was proud to look the part of a Ranger. Great-Grandpa George had served as a Texas Ranger and a soldier in the Civil War and Granddad George had served as a state representative. His father and uncles had served in WWI. The Carroll men had done their part to serve their country and Bob was determined to do his. If he completed the requisite four years of training and home study courses, he would be eligible for a commission in the Officers' Reserve Corps. Bob had set his sights on nothing less. Tomorrow, June 1, 1932, he would wave good-bye to Arthur and Myrtle and Art and Harold at the station. He would step on the train bound for Fort Logan, Colorado and into his future—even though today he was thinking of the past.

The boys turned the photo album to the next page. Staring back were George Washington and Nancy Abigail Carroll both wearing fine clothing and stern expressions.

Bob still remembered his great-grandmother Abby, Granddad George's mother, as well on in years but active, with her once raven hair now white and her voice soft and raspy. She'd remained spry and sharp-witted until her death seven years ago in 1925. She'd lived through the Civil War and was an early western homesteader. Her quiet, yet commanding words and the vision of her curling finger demanding his attention, could never be forgotten. Now, as he prepared to assume the responsibilities of a citizen soldier, her favorite saying reminded him that, "Those who can, should."

Abby talked often of civil unrest, people up in arms. There had to be "somebody able and willing to calm them down," she insisted, someone who possessed that combination of strength and reason complemented by persuasiveness. Regardless of whether conflict was in your own back yard or someone else's, "Fundamental liberties require individual responsibility!" That was Abby's contention. She'd read it somewhere and repeated it often.

"I was twenty years old in 1857," she told Bob. Her shiny black hair and flashing eyes had caught the attention of the debonair twenty-five year old George Carroll, who was the handsome and educated son of a wealthy rancher and cattleman.

"George was the son of a Georgia planter—he was smart, he was funny, and he promised to build me a home on the most beautiful land in the Guadeloupe river valley," she laughed. "But my daddy already owned some of the most beautiful land in the Guadeloupe river valley!" Fortunately George's father, Jacob, was in cahoots with Abby's father, William, both in favor of the combination.

Abby remembered her wedding day down to something old (her great-grandmother Catherine Moon's locket), and something blue (a silk garter on her shapely left leg, mid-thigh).

They gathered in the chapel in Port Lavaca; George stood tall and uncomfortable in his starched tie and top hat, (his hat and her dress now stored safely in tissue in a box in her armoire). George's parents, Jacob and Sallie, were puffed with pride and aristocratic in their bearing, both wearing serious expressions. George's older brothers—unmarried Dennis and Louis, partially cleaned up from a recent cattle drive, acted as best men. And younger brother, Jake Jr., attended with his wife, Prudie, treating her deferentially as she was in a delicate condition, pregnant with their first child. George's older sister, Mary, arrived trailing giggling Elizabeth and Sarah, the younger sisters, behind her. They were busy planning their own weddings in excited whispers, although they still needed grooms. And they had young Columbus, the little brother, in tow, a frog in his left pocket, and his hair determinedly out of place despite the many combings that were now a form of punishment.

Abby's family were a bit less refined and led by her father, William McElroy, who had been alone since her mother's death two years ago. He was still handsome in his mourning clothes and some of the town's women made a spectacle of themselves for his attention. Abby's brother, Robert, towered over the group, wearing a tall hat shadowing a long jaw covered in auburn colored mutton-chops. He was often outspoken and prone to antics. "A cut-up," Abby said, and certainly never one to miss an opportunity to show off for the ladies. And there was young John McElroy, smaller and compact, tough and ready, a bit of a braggart, though some of the talk about him was earned. John had his new wife, Elizabeth, on his arm. "She wore a bright yellow gown," Abby recalled, "with tiny pearl buttons."

It was a typical Texas wedding. Folks gathered together from far-distant places—ladies happy for an event to display their seldom worn finery, men happy for a change of pace, children happy because food of all sorts in great abundance was spread on tables under the elm trees near the church, and very little trouble was taken to supervise them.

Finally, they settled down for the formal ceremony where Abby, dressed in white lace and blushing, was presented by her father, William McElroy, to her new protector, George Washington Carroll.

Daddy McElroy had killed the big rooster and several chickens were roasted and there were baked goods and preserves, even popcorn and molasses rolled up in balls. The festivities ended with fiddling and dancing, and a fight, although by then, the happy couple had long since departed.

Alcohol and firearms were collected before the festivities began to curtail unwanted incidents. Rowdiness could not be completely dissuaded, but it could be mitigated. Folks were generally well behaved with the real rabble-rousers arriving late anyway. Bird and Jenny Kelly and John and Cora Smothers joined John and Lizzy McElroy; someone spiked the punch and fun turned to frolic.

George and Abby left early the next morning by boat to Galveston Bay, and by schooner across the gulf to New Orleans and up the Mississippi river to St. Louis, and then by train and stage to Chicago.

"Chicago was a dream!" Abby exclaimed, her eyes glistening, "Even more marvelous than St. Louis. We had our photographs taken at the McVickers Theater Building downtown. It took some encouragement to get George presentable." She laughed softly, a sound like sandpaper on a glove.

In the photograph, Abby stares straight ahead, a petite woman with good posture. She faces the camera stoically. Her dark, shiny coiffed hair parted in the middle and pulled back from her pretty face, hangs in ringlets around her shoulders. She doesn't smile. No one could hold a smile for that long. It took ages for the picture to set. It was hard enough just to sit still and hold a pose, especially one as polished as Nancy Abigail's, as if she'd practiced with a ruler, or spent many hours carrying books atop her head. Her eyes are large and bright, reflecting intelligence, and her eyebrows nicely arched. Inch long earrings dangle from her lobes and she wears a choker of braided beads attached in front with some ornate clasp. Her dress exposes her long neck; dark velvet trim runs straight across her shoulders and bust, and centered between her breasts is a silk rose.

George looks straight at the camera, handsome and proud, with broad cheekbones, chiseled features, axe-shaped nose, dagger eyes. His hair is blond, although that's not obvious from the old black and white photo, and parted on the side and combed back away from his face, hanging just below his ears. He wears the thick bow tie of the times tied around his white starched collar held tight on his broad neck, and a vest and heavy overcoat.

"Abraham Lincoln had just visited Chicago in February and the talk was all about him. We stayed three weeks, long enough for me to know I could never live in such a large city. There were so many people down-trodden and out of work, young women and children begging. It was terrible. They later called it the Panic of '57. I was very glad to return to Texas."

They traveled home from St. Louis in wagons headed to the Old Camino Real, the route that linked Natchitoches with San Antonio, and followed the trail many white Americans had taken into Texas with their slaves. George was wary as he escorted Abby home from their honeymoon, even though they were accompanied by a party of rough and ready cowboys bound for Galveston. Texas had experienced little peace since European explorers had set foot on her solid land yet ever shifting borders. Not only was he transporting his precious wife, he also had the wagons and mules loaded with flour, sorghum, corn meal, and other supplies the family was counting on to help them through the seemingly never-ending drought of this year, 1857.

He was pleased to see James McBride and stepped down to exchange handshakes and greetings. McBride operated the ferry across the San Marcos River.

"That'll be sixty cents," McBride said matter-of-factly when their wagon had safely reached the other bank.

"It was only forty cents when I saw you last. Did you raise your rates on me?" George queried.

A blustering McBride answered, "You know what I charge!" Ferry charges varied, but McBride was a stickler for routine and charged an uncompromising rate of fifty cents for a four horse stage, forty cents for an empty wagon, sixty cents for a loaded wagon with one yoke of oxen, and ten cents more for every extra yoke.

A one horse buggy was twenty-five cents, a man and horse ten cents.

"You'd charge me five cents just to stand here gabbin' with you, wouldn't you?" George taunted.

"That's right, Mr. Carroll, just like I charge a cow to cross!" Both men laughed and slapped each other on the back. After George paid, they were on their way again.

After a long and dusty ride, the wagon finally climbed up a small rise and they were suddenly overlooking the Guadeloupe river valley. There in the distance Abby saw the carpet of golden fields rolling up to the majestic plantation house, regal with its many gardens, and visible even miles away. Slavery was the only eyesore on an otherwise perfect landscape, she thought. Perhaps others had grown to ignore the pained expressions and songs of sorrow and saw only the beauty of the valley, but Abby saw a people in transition, just as she was in transition, and as the South was in transition. Many of them were in torment, separated from their loved ones as they were, and in bondage.

As they turned up the long drive leading to the house, two large red Irish Setters ambled toward them across the lawn. Running with them, was a boy of about thirteen, with hair the same color as the Irish setters, and a great grin on his face. "George!" he called loudly, racing to the wagon.

"Columbus!" George answered, "the rascal!" He slowed the wagon to a stop, leaped out, and grabbed his little brother up in a bear hug. Knuckle rubbing his setter-red head, he set the boy down. "Jump in!" he commanded.

They rode through a stately row of trees towards the plantation house gleaming white in the afternoon sunlight. Several people rested in the shade of the wide front porch. Before the wagon had stopped, Elizabeth was rushing to greet her older brother. George's older sister, Mary, waved from the porch and offered refreshments. Dennis and Jake Jr. looked up from their chess game.

Rebecca and York, the only ones not visibly bothered by the heat of the midday Texas sun, began unloading the wagon. Rebecca wore a colorful purple and gold turban wrapped around her tightly-braided black hair, and skirts and petticoats of a practical cut and material, that were kept clean and well mended. Her feet were bare. York dressed and acted like any other cowboy, the only distinction his black skin the color of obsidian. After George and Abby removed their personal belongings, York took the wagon out back to store the supplies; his young son, Barton cared for the horses.

"Welcome home, little brother. You still have your scalp, I see." The greeting from Dennis referred to George's first cattle drive and a close encounter with an angry group of Comanche counting coup. George had escaped with a bleeding

wound at the base of his skull that left a nasty two inch scar where no hair would grow.

A friendly punch was exchanged between brothers. Lizzy smiled and offered Abby a cool drink while the men continued to banter. "Prudie's not well," she confided. "She's stayed in bed all day today." Prudie was Jake's wife and just the same age as Abby, twenty-one. She was having a difficult pregnancy.

Although the siblings were playful and friendly with each other, things were unsettled from the beginning. Tension brewed between family members for control of the plantation. And tension brewed everywhere over control of the South. And tensions were often expressed during heated political discussions—talk of seceding and war.

George and Abby's arrival had provided a welcome interruption to one such heated discussion going on over the chess game, which resumed almost immediately after they'd settled with their lemonade.

"G.W. Fly is right. The North only intends to better the North. It's the rights of Texans and of Southerners we should be supporting!" exclaimed Jacob Jr. loudly. He was the youngest son, next to Columbus, and prone to assert himself forcefully where he needn't. Though now in his twenties, Jake Jr. still had a baby face, surrounded by thick, unruly blond hair. His blue eyes were passionate as he rallied to exalt his hero.

"Whoa, there, little brother," Dennis took the reins, "I served with G.W. in the mounted. He's a good man, but a bit hot headed. I'd trust him in a saddle by my side up against some Comanche, but let's see if we can't take a couple of steps sideways to avoid a war with our neighbors instead of taking the plunge right straight into the thing."

Jake Jr. scooted his chair back abruptly, enjoying the grating sound of the scraping chair legs against loudly resistant porch boards, and made his comment without words. He dipped his hat politely, if curtly, to the ladies and loped off in the general direction of the barn.

George Washington Carroll, Chicago, 1857

CHAPTER 9

ANTEBELLUM COWBOY

When the first gun is fired away goes all your slaves. They will be freed and you will lose millions of dollars now invested in them. The rivers will run red with the blood of the best citizenship of the country. Owls and bats will take up their abode in your houses and grass will grow green in your streets.
~ Sam Houston, first Governor of Texas

Just after George and Abby married, the U.S. Camel Corp traveled from San Antonio to El Paso with a procession of troops, mules, horses, civilians, and 75 camels.[39]
Many turned and gawked at the sight of the first thirty-three camels that were unloaded at the Texas port of Indianola on the bright, sunny morning of February 10, 1856. Dinner tables were buzzing with gossip about the strange creatures that evening. The next year another forty-one camels joined the original group and their caretaker, "a short, heavyset, happy-go-lucky Arab named Hadji Ali." Hadji Ali was quickly renamed "Hi Jolly" by the soldiers and local cowboys.

The camels were based at Camp Verde, Texas as part of a failed U.S. military experiment. "I'll tell you why the government quit fooling with the camels," Bill Keiser told William Barnard of the associated press. "It wasn't because they didn't do a good job. They could carry a thousand pounds of freight 65 miles a day and they went three days without water. But they scared hell out of every varmint that sighted 'em and they caused plenty of trouble.

"Hi Jolly told me all about it. Those camels were lonesome for the caravans of their home country and every time they sighted a prospector's mule train they'd make a break for it. You've heard of how horses bolted at the sight of the first automobiles. That wasn't anything compared to the fright those ugly, loping camels threw into mules. The mules would lay back their ears and run for their lives and then the prospectors would cuss and reach for their guns and shoot at the camels. A lot of camels got killed that way." Bill claims camels still roam the southwest.[40]

Although camels failed to establish roots in Antebellum Texas, slavery flourished. Jacob Carroll was a man of his time and place and, like his contemporaries, saw his role as planter and rancher as being integral to the survival and growth of the South, and of the life that he held dear. He knew that plantations like his provided the raw material from which the North produced its dirty, industrial wealth. Jacob preferred his agricultural livelihood, and felt that slavery was a natural part of a working system that played a crucial role in the economy as a whole.*

Jacob Carroll might have asserted that the life of a slave was better than that of a poor working man stuck in the labor mills or sweat shops up north; slaveholders, unlike greedy northern employers, often provided for their workers when they were too sick or old to contribute. In fact, census records show that a 90 year old black man by the name of Nero resided with daughter Sallie Carroll and husband James Turk and their young family in 1870, after the end of the Civil War.†

While he might try to assert otherwise, Jacob could not well defend the accusation that the South was backward. The South languished behind the North because slavery "retarded the rise of industry and commerce. Slave labor made the plantation productive and profitable and reduced the need for the invention and manufacture of farm machinery."[41]

For successful planters like Jacob Carroll, there was no motivation to support a change, especially a change that might end slavery. Instead, there was motivation to fight for the status quo. Jacob didn't want owls and bats living in his home or prairie grass growing in the streets of Gonzales.

Few Texans owned slaves, but some Texans had grown dependent upon them in order to keep their wallets fat. In 1850, the Gonzales County population was 1,492 and 601 slaves. In 1860, Gonzales County population was 8,059 with 384 slaveholders holding 3,168 slaves. On February 23,

---

* This is not the author's view.

† While this view has some merit – some slaves may have had better healthcare, for example, than freed African-Americans – it fails to acknowledge the tragedy of slavery. Some slaves were beaten and murdered.

1861, with the encouragement of Governor Sam Houston, residents voted 802 in favor of secession from the United States and 80 voted against. The voices of those who might speak out against slavery were drowned out by the crowd.[42]

Jacob was the patriarch of a large household and held many slaves. The household included his eldest son, Dennis, who regularly proved himself in the local militia and kept the range safe for civilized folk. Second son, Louis, also a ranger, had been groomed by Jacob to take over the operation of the plantation and Jacob depended upon him more and more each day. Louis had taken over driving the cattle to market—transporting cattle and cotton to New Orleans and Abilene—an activity that Jacob was happy to pass on to the younger man, along with the aches and pains of the bones that went with it!

Jacob sold a portion of his land to his middle son, George Washington Carroll, to build a home for Abby and their new family, and so he'd be nearby and could continue to help work the plantation. Younger son, Jake Jr., drove cattle with George and his older brothers and gained a reputation as a skilled hand. And Jacob's youngest son, Columbus, assisted with the operations as a child and later made a name for himself as a trail blazer driving cattle from Texas to Montana with brother Jake.[43]

The ladies of the household, including mother Sallie, and her daughters Elizabeth and young Sallie, and daughters' in law Abby and Prudie, traveled thirteen miles north to the town of Gonzales for entertainment and shopping several times a year. Eldest daughter, Mary, married now to Dr. Johnson, often joined them. By 1857, Gonzales had eleven large dry goods and provisions stores, two drug stores, two furniture stores, one silversmith shop, two hotels, several boarding houses, four blacksmiths, two livery stages, three or four carpenter shops, two tailor shops, two paint shops, two bakeries, a printing office and a newspaper.[44]

"Spiritous liquors" could be purchased in quantities of less than a quart, in the house on the south side of Public Square known as the Ten Pin House next door to the Post Office. And shoppers had their choice between "moss or cotton mattresses," with "orders filled immediately" according to the Gonzales Inquirer.

Schoolhouses were rare in Antebellum Texas, but literacy was high as attested by the many newspapers distributed in Gonzales County. Abby would have been responsible for educating young George (Granddad George) and taking care of household chores and the kitchen garden.

Natural disasters plagued the county the year George and Abby married. The drought of 1857 was so severe it caused a rumble in the legislature as well as in the belly, and prompted action from the legislature. It was estimated that fifteen thousand bushels of corn would be needed to "bread the county." Later that year, grasshoppers invaded laying eggs that hatched the following spring and devoured the crop.

The Carrolls were a tight knit family in a community that was tight knit by necessity. Men of Jacob's day were expected to fight and provide. Local militias had formed to fight the Indians and defend the newly established Texas border. Strong, independent women ran the fields and households while the men were away and then inherited the property, including slaves, after the men died.[45]

Jacob's contemporary, John Smothers Sr., was a patriarch like himself. The Carrolls and the Smothers families had intermarried as had the Carrolls and Kellys, and the Carrolls and Turks. And the Carrolls and McElroys had family in common when George and Abigail tied the knot. And of course the Smothers were related to the Kellys and the Turks and the McElroys and the Karneys.

John Smother's stepson by Mary Ashby was one Charles Carroll. Charley Turk married his daughter, Lucy Smothers. E.A. Turk married her sister, Fanny Smothers. Charles McElroy married another sister, Lizzy Smothers, after she'd married and buried a Kelly. And James Turk was about to marry Jacob's daughter, Sallie. In sum, all of these powerful Texas families of Irish and Scots-Irish descent had intermarried and produced a brood of offspring.[46]

Jacob Carroll and John Smothers had strong interconnections and much in common. Both had roots—family and education—from back east, and both raised their children on the wild and isolated frontier without the aid of eastern institutions.

Jacob wasn't necessarily ready to welcome all of the Smothers with open arms. Smother's brother, William, had a reputation of ignoring social convention. Locals saw John as the more respectable of the two. William Smothers sold liquor and ran a gambling establishment in a shed behind the store in Hallettsville, Texas. Folks just shrugged when hearing the news of an incident or fight involving Bill Smothers. It wasn't news. Sometimes the judge slapped him with a fine, one cent was typical, but then he'd be back at the shack with a bottle of whiskey and the story going around the poker table.[47]

He'd faced legal charges at least nine times, and he'd paid cash for fines at least six times officially. Yet he stayed in business because Bill had earned the loyalty of many starting way back in 1840 when he, along with other Lavaca settlers, routed the Comanche at Plum Creek, and again in 1842, when he guarded against Mexican marauders as a spy in the Hays' Scout Company.[48]

Bill Smothers had even faced off with Sheriff John McKinney. That had cost him five dollars. After the Sheriff charged him with assault on

November 4, 1856, Bill complained that the local officials had it out for him. Eli Holtzclaw, the assaulted, filed charges but didn't have the clout of Smothers. And whether or not Bill was responsible for Eli's injuries, he celebrated yet again with a round of whiskey with the boys out back after he was found not guilty.[49]

Politics was the topic on everybody's mind. Unlike gambling, a man could spend his time politicking and feel like he'd done something of value at the end of the day. Any man able would happily ride half a county through the rain to the nearest polling booth or to hear a rabble-rousing speech. Who knows what sparked the change, but in 1858 Bill Smothers started rounding up volunteers, calling on those he knew during his days ranging back in the 1840's. "On May 25, 1858, on learning that Governor H. R. Runnels was organizing six companies of volunteers for frontier service, Smothers raised a company of one hundred men, 'a large number of whom are old Texans and quite familiar with ranging,' and tendered their services."[50]

The atmosphere was tense with the electricity of an imminent war. Bill watched as they signed-up the rabble-rousers, the Regulators and Moderators, and even some of the Carroll boys under the leadership of G.W. Fly. Smothers found it a natural progression to go from giving orders in the shed out back to giving orders to young recruits. The next step was to run for office.

In 1860 he was a candidate for sheriff and, despite thirty-three indictments on his record, he defeated A. G. Nolen in a hotly contested race…

According to researcher Paul Beothel, his brother, John Smothers, "was a cattleman who had earned his keep and built his pile as a freighter and stock raiser. In the 1850's, with Hallettsville as headquarters, he and brother Charles entered the freighting business handling freight to Port Lavaca, LaGrange, Austin, Fort Mason and other points."[51]

"At the outbreak of the Civil War, the commissary general of Texas was having difficulty supplying beef for the army, and Smothers was detailed to Louisiana to collect cattle. After the fall of Vicksburg and the capture of New Orleans, it was no longer feasible to transport cattle from Texas to the armies in the East. Thereupon, Smothers was ordered home to report to Captain John Pulliam at Hallettsville, who had organized a company of troops there. He remained in service with the company until July, 1863, when he hired a substitute, Patrick O'Dougherty…Smothers, once out of the army, resumed his work of gathering beef and shipping supplies to the troops in Texas and Louisiana."[52] The Smothers family and friends assisted with these cattle transporting operations, including some of the Carroll boys.

SPRING 1933

Bob's cowboy hat hung on a peg in the kitchen next to his CMTC hat, with its flat brim and secure leather chin strap. Bob had trouble deciding which hat he preferred to wear—the military hat with the honor and prestige that it signified, or his old cowboy hat with its history and the independence that it signified.

He sat with Arthur at the kitchen table, the straight-backed wooden chairs uncomfortable. The steaming coffee and warm company were inviting as the morning was still cool with a dewy breeze blowing in off of the mountains. Arthur ruminated on something or other in the paper and folded the page in half to make it easier to read. The back door was open, and they could see Myrtle through the screen door working in her garden. Flowers were in bloom.

Bob was enjoying his last Sunday at home. He was preparing to leave for Fort Logan for four weeks training at the Citizens' Military Training Camp, his second summer participating, this time in the Red Course of Instruction. He was edgy with excitement, even looking forward to the hard work. It would bring rewards if also sore muscles, though it was not like he'd been sitting around on his rear. He'd worked hard since his high school graduation as a brakeman for the Burlington Route, suffering under the scrutiny of his engineer father's railroad cronies who gave him no mercy.

"How'd it go yesterday? You got back pretty late," Arthur's disembodied voice asked from behind the paper. He finally passed the pages over for his son to read. Looking Bob in the eye, Arthur raised his eyebrows to emphasize the unanswered question.

Bob took the pages offered. "Some of those gandy dancers* are real pranksters," he complained.

"You've only worked with them a few months. Everyone has to pay his dues," Arthur reminded him. "You'll earn their respect eventually."

"Eventually, I'll have moved on and they'll still be here. I guess that's revenge enough." He took a sip of hot coffee, burning his tongue.

Father and son read over the paper together every Sunday. Bob was particularly interested in the international news. He turned to the article his dad had been mumbling over. Germany's Chancellor, Adolf Hitler, and his Nazi party had proposed a bill that would make Germany a dictatorship and Hitler the dictator. His Nazis handled those who spoke against them, mostly Communists, by gathering

---

* "Gandy Dancer" was a slang term used to describe early railroad workers.

them into concentration camps or driving them into hiding. Bribery was also an effective tool and when the Catholic Church backed the Nazi party there was no voice of power left to object to the bill empowering Hitler, and so it passed.

Hitler and the Nazis gained control of the trade unions. Then they removed the union leaders to concentration camps. Hitler's minister of the interior, Goering, replaced senior police officers with Nazi supporters, creating a brutal force called the Gestapo. Joseph Goebbels developed propaganda on a scale never seen before as Minister of Propaganda.* All information that people in Germany received was carefully selected and modified. Schools posed a real threat with the possibility of children learning ideas counter to Nazi interests. And so the Hitler Youth were invented and a new carefully crafted message was taught. The Nazis even burned textbooks if they didn't support the Nazi agenda.

Hitler was not a typical military sort. He was weak, unwell, bad-tempered and had a reputation for "eccentric behavior." Some thought he was strange. He'd been a spy for the German army and could lie with ease to achieve his goals. And people believed him. He was a convincing speaker. He read voraciously. He was influenced greatly by American Henry Ford and his book, *My Life and Work*. Ford claimed there was a Jewish conspiracy to take over the world.

Arthur mumbled concern, "Bad feeling about all this," nodding to the paper, "history repeating itself," he harrumphed as he rose from the table, leaving his son with his own thoughts to ponder. Bob knew that Arthur was off to his morning constitutional.

Bob rinsed his cup and saucer before seeking some privacy in his room at his writing table. He recorded his thoughts about his new girl, Rachel, the world, his military training and now, the importance of the freedom to learn, in a small, leather bound book of pages. He remembered Great-grandmother Abby telling him how she calmed herself in troubled times by writing in her diary, and that she had done so all her life long. People often acted in response to events in a hot headed manner, she'd lamented. If they would just wait, give it a little time, reflect on it, write about it, things would work out better in the end.

Back in 1858, just after she'd married George, Abby's diary entries had focused mostly on family and household matters—Jacob's wife, Sallie, and her protracted illness, and the impending marriage of their daughter, young Sallie, to James Turk. Abby and George were crowded into a few small rooms at the back of the second

---

* Propaganda: "Any association, systematic scheme, or concerted movement for the propagation of a particular doctrine or practice." Compact Edition of the Oxford English Dictionary, Volume II, Oxford University Press, 1971.

level. Plans for their own home hadn't yet materialized. Abby watched the days, weeks, and months tick slowly by on the grandfather clock in the entryway. Finally, in 1859, just before the war broke out, Abby stood, big with child, on her own front porch overlooking the valley.

Her diary pages reflected moments of fleeting joy—private soirees with George at the site they'd selected for their new home, parties with neighboring families that sometimes included a hired band and dancing, and the birth of their son, George Jr. [Granddad George]. Her pages also reflected the truth that life was often painful and short. A disheartened townswoman she'd talked with last month in the dry-goods store described to her in a measured monotone how her father, mother, four sisters, three brothers, and two infants had all died in the past two years. Except for her father, none had reached the age of forty.

In order to keep her mind nimble and remain pure of heart, and to help with her children's education when the time came, Abby read voraciously every book that caught her interest in the Carroll library. Currently she was enjoying Milton's Paradise Lost. *Although she might reprimand herself for engaging in such idleness, as opposed to reading* The Decline and Fall of the Roman Empire *by Gibbon, she justified her light reading as a rare pleasure for a hardworking wife and mother.*

Abby's dear friend, Henrietta, was attending Gonzales College and hoped to earn her A.B. degree. * *Abby much admired those women with fortitude and ability to earn a formal education. She had a passion for learning herself. But an increase in household duties due to the birth of George Jr., the more frequent absence of her husband, and her sense of impending disaster prevented her from anything more than a vain attempt at nightly scribbles or reading a novel.*

George never resented Abby's desire to increase her knowledge of the world. He even respected it. And Abby trusted and admired her husband. George Washington Carroll was a calm and gentle man, secure in himself, not a boastful talker, but a man of dignity who could be hard or tough if he had to, but never mean or petty. Although sometimes Abby saw George, and the entire race of men as selfish louts, unable to see beyond a foolish and narrow vision of the world, a vision that typically excluded women. How could war help to make any of this better? She wondered. Abby found the whole situation maddening.

More exasperating was George's increased time spent away from home that started before the war began. The militia groups had congregated in the kitchen

---

* Gonzales College was founded in 1851 by slave-owning planters and was the first institution in Texas to confer A.B. degrees on women before the Civil War. It is a surprising contradiction of prejudice that women's education was considered important by slave owners.

and ranch house for years. But things had grown more organized and intent. G.W. Fly was now gathering a small group of volunteers in Gonzales County. These men included three of the Carroll boys, of which Jake Jr. had become the most ardent spokesman and recruiter. When they named themselves the Gonzales Invincibles, Abby knew the trouble was just beginning.

Nancy Abigail McElroy Carroll, Chicago, 1857

CHAPTER 10

THE GONZALES INVINCIBLES COMPANY
It was only a short time until I had strong suspicions that I had joined a regiment of devils. ~ Private Ralph Smith, Company K, 2nd Texas, 1861

THE OLD RANGERS STARTED TO CONGREGATE. THAT WAS A CLEAR SIGN TROUBLE WAS IMMINENT. BILL SMOTHERS gathered his group of rowdies, gave them guns and called them volunteers. And when George Washington Fly gathered his volunteers, they included the Carroll boys.

According to descendant Betty D. Fly and Craig H. Roell, "Fly was a staunch supporter of states' rights and a regionally noted orator. He favored the Breckinridge-Lane ticket in 1860. During the Civil War G. W., as he was called, was a seasoned commander in the Second Texas Infantry and commandant of Galveston.

"In 1861 he gathered a small group of volunteers in Gonzales County who elected him their captain. These men were mustered into Confederate service as Company I, Second Texas Infantry, known as the Gonzales Invincibles, and later joined the Wilson Rifles to form a complete infantry company. Though designated the second, this unit was really the first infantry regiment organized in the state. Its colonel was John Creed Moore."[53]

Before the Civil War was even declared, Fly recruited mounted riflemen to serve in the Gonzales Invincibles. George joined with his brothers Louis and fervent Jake Jr. And his sister Sallie's husband, Louis Turk, and his brother, Charles Turk, also enlisted. The Carroll & Turk boys rallied around their leader, G.W. Lafayette Fly.

The "Gonzales Invincibles" were also called the 2nd Sharpshooters.

The roster lists 1,059 soldiers of the ten companies:[54]

Company A - (The San Jacinto Guards) Houston, Harris County
Company B - (The Confederate Guards) Houston, Harris County
Company C - (The Bayland Guards) Houston, Harris County
Company D - (The Confederate Grays) Houston, Harris County
Company E - Robertson and Brazos counties
Company F - German immigrants from Galveston
Company G - (The Burleson Guards) Caldwell, Burleson County
Company H - (The Lexington Grays) Burleson and Lee Counties
Company I - (The Wilson Rifles) (The Gonzales Invincibles) Gonzales County
Company K - (The Texas Guards) Jackson County

When war was declared, Texas men and boys, and even some women, volunteered for military duty. Gonzales, Texas went from a rough and rowdy, but fairly prosperous frontier town to a fairly dangerous place in a short span of time. Reins that had always been loosely held were released and those who might take advantage of the situation did. "The political system was in great upheaval during the Civil War. There were no nominating conventions held and all candidates were self announced."[55]

As the war continued, the number of men in Gonzales County shrank. The women stepped forward and shouldered the responsibilities as best they could. Finally, out of frustration the townswomen created a "Spinning Wheel" petition, so called because the women who signed it did so in the form of a wheel, without dating it. It read: "We, the undersigned petitioners would respectfully ask of your Honorable body to have Thomas Baker detailed to remain at home as our neighborhood is being left entirely destitute of men, he being calculated to do much good, particularly in making looms and coffins and other work in that line and has already proved to be of much benefit to our neighborhood in butchering both beeves and pork." This home-spun petition was signed by: Mary Jenkins, Margaret A. Kelley, Mildred Clark, Nancy Baker, Malinda Power, Louisa Matthews, and Eliza Gates.[56]

News of missing husbands, brothers and sons came sporadically, if at all. Some loved ones just never came home again.

G.W. Fly served in the battles of Shiloh and Iuka. The military sent notice to his family that he'd died in battle at Corinth but, in fact, he was captured, exchanged, and finally returned to his post. Before the siege of Vicksburg, he was promoted to major and in August of 1864 he was made commandant of the post at Galveston. He defended Galveston until the war ended and then returned to his family in Gonzales County. The Gonzales Invincibles formally surrendered in May 1865.

The Carroll Boys returned home to their plantation—to abandoned cotton fields and a cattle business destroyed by the war. But they were resourceful; there were a lot of unbranded long-horns running wild in Texas, free for the taking. Jake Jr., Columbus, and their brother-in-law, Charley Turk are recorded as among the first to drive cattle to the newly opened markets in Kansas and up into Montana in search of a livelihood after the war. They likely took their best hands with them, including York's son, and other former slaves from the Carroll plantation.

Freed slaves and other skilled ranch hands found work doing what they knew best after the Civil War ended. Black cowboys ranched between the Sabine and Guadalupe rivers. Many found employment breaking horses and branding. Some became ranch foremen. According to W. T. Block, "after the Civil War, ranches east of the Trinity River often had all-black crews. West of the Nueces River, ranchers employed vaqueros more often than black cowboys. Far fewer blacks populated the northern ranges. Montana censuses counted only 183 blacks in 1870 and 346 in 1880."[57]

"The trail drives north from Texas (1866 to 1895) employed about 63 percent white, 25 percent black, and 12 percent Mexican or Mexican-American cowboys. Unfortunately, most black and Hispanic cowboys faced social and economic discrimination in the West as they did elsewhere in the country."[58]

Some black cowboys became Federal officers providing safe passage through Indian Territory for the intrepid, yet unprepared white pioneer. Others became gunfighters and cattle thieves joining a multicultural group of Western outlaws.

Some joined the rodeo showing off their best riding and roping skills including William Pickett, one of the most well-known Wild West rodeo cowboys in the country. Pickett was a steer-wrestler. Called "bulldogging," Pickett wrestled steers and worked on Texas ranches during the 1880s and 1890s. He later started the Pickett Brothers Broncho Busters and Rough Riders Association. Pickett died at age 101 in 1932 after being kicked by a horse. He was the first black cowboy admitted to the National Cowboy Hall of Fame in 1971.

Many Black cowboys made a name for themselves including Bose Ikard, Charles Goodnight's ranch hand, who was portrayed by Danny Glover in the 1989 Lonesome Dove miniseries.

After the Civil War ended George and Abby gathered their children and possessions, and then they left Gonzales and cattle ranching and moved to Bell County, Texas for a change of scenery. There in "hill country" the Carrolls built a mill and raised their children.

From genealogical records compiled by Merle Harris we learn that when emancipation was proclaimed, "George, Sr. was the owner of a very large number of slaves and…financial difficulties caused them to leave Gonzales County and move to a new location." Records also state that in or around 1861, George, Sr.'s health failed. He never completely recovered. As a fighting member of the Gonzales Invincibles Company, he had engaged in action, possibly at Shiloh, early in the Civil War. His health "failed," in or around 1861, likely as a result of his service as a mounted rifleman. Possibly due to a gunshot wound.

Bob, Ruth, & Rachel, mid 1930's

SUMMER 1934

The air was crisp and the breeze cool although it was the peak of summer in the Bighorn Mountains. Bob wore only a light jacket of smooth, roan colored leather—a gift from his Uncle Robert E. last Christmas that he cherished. He sat tall in the saddle atop the fifteen-hand palomino. He looked fair and regal, Norse like, with every hair in place, mostly because he'd just left training at Fort Logan and his military cut hadn't grown out yet.

Rachel snapped the photograph when he was unaware, not looking or smiling at the camera. But he smiled when he saw her and galloped over to the Ford Coupe parked in the shade of a pine grove, a picnic basket on the hood. Rachel and the smell of fried chicken beckoned. Her best friend, Ruth, was there with her. They couldn't be separated, "attached at the hip," he accused whenever he saw them.

The girls spread out a blanket over a grassy spot fragrant with pine needles and displayed the bounty: fried chicken, potato salad and peach pie, even a thermos of lukewarm coffee, though Bob would have preferred milk.

"Where's Ron?" Bob asked Ruth, as they usually went out as a foursome.

Ruth reached for a drumstick. "He's working in Cody for the Corps," she said between bites. "You wouldn't believe the relief on his mother's face when she got the first twenty-five dollar check. She can't say enough positive things about Ron and the CCC. When did you get back? Did you have fun?"

"Fun? I don't know if I'd call it fun, but I got home Sunday." Bob relayed some of his adventures at the military training camp, skipping any stories the women might not enjoy. He'd completed the White and Red Courses of Instruction, drilling the previous two summers. He had only one drill left to complete and then he would become a commissioned officer.

After lunch, Bob and Rachel leaned against an Aspen tree, his legs stretched out and crossed at the ankles. Rachel's head rested on his shoulder as they talked.

Rachel squeezed his bicep. His hazel-

Bob & Rachel, mid 1930's

blue eyes, now the color of robins' eggs, sparkled with mirth. He wrapped his arms around her in a passionate embrace.

"All that work's paying off, I guess." He smiled and kissed her. Bob was filling out, gaining weight in muscle mass. He had been willing to hang by his fingertips until they were numb in an attempt to make himself taller. He'd go to even greater lengths to ensure his acceptance as a commissioned officer. He stayed focused on his studies, at least most of the time, and had passed all of his exams with a 90% or higher, usually a 98% or higher. His goal was perfection. Bob knew he was destined for a great military career. He was no "Tommy Baker". Unlike Adolf Hitler, who in spite of his leadership, was in Bob's estimation just like the Tommy Baker his great-grandmother Abby had described with derision.

Bob couldn't help discussing politics, raised on it as he was. "Hitler has no original ideas," Bob denounced the new German Chancellor. "He copies us." Hitler had pointedly studied Roosevelt's Civilian Conservation Corps and imitated the idea, sending German boys of the "voluntary" Youth Service out to plant trees, repair riverbanks, and learn about land reclamation.

Hitler's Youth also learned the Nazi message: those of Jewish descent must beware. Jewish teachers in Germany were now jobless. Signs appeared. They bore different words but had a similar message to signs in the America's South: "Jews not admitted." The intent was sinister and systematic. Segregation was not the goal, as it was in America. The goal was extermination.

"Hitler's policies are forcing out Germany's best men. Those who can are leaving and seeking refuge here.* His is an arrogance born of insecurity. His decisions come from weakness not strength. He's just a Tommy Baker."

"Tommy Baker? Who's Tommy Baker?" Ruth asked, puzzled.

"Tommy was a real palooka," Bob answered cryptically.

"But how is Hitler like Tommy Baker?" Rachel stretched out her long blue-jean clad legs and rested her head in his lap, directing the question to the sky.

"Hitler, like Tommy Baker, is a below-average social outcast full of resentment for his shortcomings, but I'll take Tommy's brand of resentment any day. Hitler seems a nasty sort." Bob started whistling the sad end of a Duke Ellington tune, running his fingers through Rachel's blond curls. His great-grandmother Abby's vivid tales of Tommy Baker and the Civil War filled his imagination.

Bob, Art and Harold had grown up listening to stories about poor Tommy Baker. Whenever a child was left out of some event because he was too small or

---

* Almost half of the population of Germany sought new lives in other countries between 1933 and 1939.

too young, Great-grandmother Abby referred to him as a Tommy Baker.

"Just like poor Tommy," she'd say.

Tommy was unwillingly left home and out of the Civil War because of his congenital deformity; one leg was almost an inch shorter than the other, causing a decided limp, and because of his young age. Tommy Baker was only fifteen in 1861. But he still wished for the glory of battle and dreamed of escaping the clutches of the women of Gonzales, Texas.

The day before his sixteenth birthday on June 11, 1862, early in the morning, Casey's chickens found Tommy crouched behind the barrels of oil in back of Casey's dry goods store. He was hiding from Mary Jenkins, Margaret Kelley, Mildred Clark, Nancy Baker, Melinda Power, Louisa Matthews, and Eliza Gates. The Spinning Wheel Women were making his life a living hell.

Hardships began shortly after the Civil war started. The shops and farm stands in and around Gonzales, once plentiful and well stocked, dried up like a summer creek. There was little or no access to things much needed and depended upon, including men. The once prosperous life of the plantation grew miserly. The strain of the men being gone and news of the wounds of war, brothers and sons and husbands returning without limbs and eyes or worse, caused beautiful women to wear dark circles proudly. It also caused some to pursue poor Thomas Baker, the last man in town, like he was the last man on earth.

Rural life with its seasons of planting and reaping helped keep the time, and marked the passing of often very boring days in which almost any bit of news, any adolescent prank, any gossip about romance helped break the monotony. The women tried to keep the gossip about hardship to a minimum. It could be their own sorrow talked about tomorrow. A pleasant way to mark the time was to pursue Tommy Baker. Ostensibly this was to lift or haul something heavy, or perform some other task only his manly frame could manage. And "poor Tommy Baker" was always an acceptable topic of conversation, if undignified, giving the women something besides war to discuss.

Or one could be dignified. The always proper Martha, now widowed, had taken to drink to ease her burdens. Ellen was doing the housekeeping for several families to maintain her own home and family, and she was looking haggard and could be heard scolding her children frequently. Abby knew she should never complain about her own problems; she must accept the burden she carried and be grateful it wasn't more. If to ease this burden, she chatted a bit too often with others about a certain Thomas Baker and his business. Well, what harm was there in that?

In addition to supervising the work of household slaves, Abby worked very hard herself. She cared for children—hers and others—helped care for pigs and chickens, planted and harvested both summer and winter gardens, sewed clothing for herself and little George, Jr., and others in the household, and cooked and cleaned. And that was before the men left to fight the war.

Abby had been raised to pity and fear the Africans as beneath their masters. And yet she could not help but admire their resilience in the face of adversity. The Civil War had left them all, slave and master alike, doing without. Yet the Africans still smiled at their children, sang their sad songs in the fields, and entertained each other with music and dance, even when weary from work. They seemed to grow strong from adversity while many whites around her broke from the strain.

Slaves on the Carroll plantation were ranch hands and cowboys. York and his son, Barton, had always been polite and respectful since first introduced to Abby. As time went on, George and the other men returned less frequently and for shorter periods of time, and Abby grew to depend more and more on the strength of a people she was raised to look down upon.

An old African woman, originally from New Orleans, was called upon frequently for difficult childbirths as the male doctor had disappeared months ago. Certain boundaries seemed to disappear in times of trouble. When Abby found one of the Carroll's household slaves, Melissa, and some small children placing clear, shiny rocks around the house, she inquired as to their nature and properties. She had never seen rocks like these before.

"They're crystals, ma'am," Melissa answered with an air of mystery. "They'll help to keep us safe."

Abby nodded, intrigued. "May I have one?" she asked. Melissa smiled and Abby reached into her upturned palm. She caressed the smooth surface of the many sided stone and thanked the girl, leaving her to her strange, protective duties.[59]

Rumors of the war, including the Battle of Shiloh on April 30, 1862, reached them before any official word. In fact, the official word never came at all. One day Louis rode up leading a horse with George hanging haphazardly from the saddle. Louis' expression was solemn and Abby and the others ran to help.

"He's been shot," Louis stated. "Let's get him into the house." George slid from the horse and accepted assistance from York and Barton. His ribs had been crushed by the horse that, also wounded, fell on top of him, serendipitously protecting him from certain death.

Abby and Sally cleaned and redressed Georges' wounds; the bullet entrance was surrounded by angry red, swollen skin. After some rest, he told of the battle, with Columbus hanging on every word, while Louis paced restlessly near the door. His mind was obviously elsewhere, mostly on the battlefield that he was hell bent to return to.

George felt his restlessness. "I'll be fine," he assured Abby, "I'll just need some time to heal. Then I'm going to find that dirty buck skin scoundrel." He coughed weakly, his will overpowered by his pain. She knew he was home for good. The bullet had lodged so closely to his lung that the doctor was unable to extract it. Louis told her he'd recover but never completely.

At least the tension had been broken for Abby. The tension of waiting for word from the man she loved, waiting to know that he was safe, that he was alive, and that he was coming home. She hadn't realized until now just how tightly she'd held her shoulders. The muscles in her neck had strained against the fear of his death, against the pain of loss, and now finally she felt the taut string slacken.

Abby's trust and belief in God had been tried to the limit. Where was God in this destruction? Everything about life had to be redefined. She finally found her faith again renewed with his safe return, and she thanked God.

Nellie Carroll Thorn

CHAPTER 11

MAVERICKS

A man named Maverick was a large rancher prior to the Civil War, and allowed his calf crop to go unbranded during the later years of the war. At the close of the war there were thousands of his cattle without brands. Therefore, in Maverick's section, folks would say, 'there is a Maverick', when referring to an unbranded critter. The term was taken up by others, and in a short time it was in general use through the range country of Texas. ~ A. M. Garrett

GEORGE, SR. SETTLED HIS FAMILY IN BELL COUNTY, TEXAS AFTER THE CIVIL WAR ENDED. GEORGE AND ABBY purchased a farm and built a flour mill and cotton gin. At this time they had the distinction of owning the only flour mill and cotton gin in the northern part of Texas.[60]

They lived comfortably and raised their young children in beautiful hill country. Granddad George (George Granville Carroll)—named after the English Earl of Granville—grew strong and tall and became responsible at a tender age in order to assist his father, who still suffered bouts of poor health. Granddad George's younger sister, Lillian, five years his junior and born just after the war ended, was also tall and tough, as well as beautiful. His youngest sister, Nellie, was almost ten years his junior and destined for a colorful future. And Granddad Georges' younger brothers, Fred and Solon, were just twinkles in their father's eye.[61]

The close of the Civil War did not end strife and discontent in Texas or the South. The Reconstruction, in fact, added to the continuing hardships. Anger and fighting persisted, fueled by anti-Union or anti-Confederate sentiment. Bell County was fired-up and embroiled in the Early-Hasley feud that lasted from 1865 to at least 1869, and drew anyone with an opinion from miles around like flies to honey.

Townsfolk said that John Early had beaten Drew Hasley, who was an old man. John Early was a member of the Yankee Home Guard and when Hasley's son, Sam, returned from Confederate service, he sought revenge. Yankee officials and others supported John Early. Hasley's supporters were mostly former Confederates. The Carrolls may have had difficulty staying out of the fray.

Family records tell us the Carrolls moved away from Bell County due to George, Sr.'s failing health, but the real reason might have had to do with the Early-Hasley feud. The Carrolls moved, coincidentally, after the flour mill and cotton gin they'd operated since the mid 1860's was burned to the ground in 1872, in what was likely a deliberate act, or at least without any recorded explanation. George, Sr. and Abby packed up their belongings once again. They settled this time in Young county, Texas and returned to the cattle-ranching business.

George, Sr. may have been encouraged to return to ranching by his old Civil War Commander, G.W. Fly, from the Gonzales Invincibles. Fly was now living along the Brazos and raising cattle. The Carroll family moved nearby and homesteaded land in what much later became a large oil field known as South Bend on the Brazoa River where they cattle-ranched for most of the next decade.[62]

George, Sr.'s younger brothers, Jake Jr. and Columbus, also returned to cattle ranching. And the Smothers brothers continued raising beeves, as they'd never stopped. They'd supplied the army right through the Civil War.

There were many new comers who tried to raise a herd after the war ended. A.M. Garrett of Shelby, Texas explained to his grandchildren how Mr. Maverick's name had become a term for "a wild one". Garrett said:

> Many others neglected to brand and mark their cattle. The reason they neglected their herds was because after the first year of the war Texas became isolated from all Confederate states east of the Mississippi River.
>
> Being cut off from the market, the ranchers found themselves with worthless stock, so far as a market was concerned. In fact, the value of cattle, in Texas, was so low one would lose money paying hired help to attend a herd. Therefore, the ranchers gave very little, if any, attention to their herds. The herds multiplied rapidly. Thus, when the war ceased, there were thousands of unbranded cattle over all the range, and no one knew to whom the animals belonged.
>
> Immediately following the close of the war, there existed a demand for beef, but Texas did not have an adequate market. Her Southern market was depressed, because of the financial condition of the Southern States at the time. There was no money in the South and the prices low. During the late 60s and early 70s,

> railroads built railroad lines down into Kansas and that placed shipping facilities close enough to our range so it was practical to drive herds to the railroads." Following the completion of the railroads extension into Kansas, cattle market centers opened up.
>
> When these markets were established, demands and prices for cattle multiplied in a short time. It is obvious to anyone, that with the range being open, it was impossible for a person to identify any particular unbranded critter as belonging to him. Because of this fact, there took form a sort of gentlemen's agreement to govern the branding of the Mavericks, and the only logical rule, ... was that a rancher had the privilege of branding the unbranded cattle which were found with his critters or grazing on the range under his control. There were many persons who started a herd by the simple process of locating a watering place, adopting a brand and then going out on the range to hunt and brand Mavericks.[63]

Life on the range became profitable once again. Longhorn cattle had survived on their own while Texas ranchers went off to fight for the Confederacy. Untended, their herds had grown. It was just a matter of gathering them up, fattening them up, and getting them to market.

Tight-knit associations with other ranching families helped the Carrolls succeed. The Carrolls, Turks and Smothers families remained connected. According to Paul C. Boethel, "By 1885 Smothers and his wife, Mary, had a brood. . . . There were four sons, Robert, Henry, John W. and George; and five daughters, Lucy, Maggie, Fannie, Jennie and Lizzie. They had extensive land holdings in the southern part of the county, stretching from an area west of the Lavaca River east to the Navidad.

"Their brood had grown with the advent of in-laws. There was quite a contrast among the in-laws. Lucy, who married C. C. Turk; Fannie, who married E. A. Turk; and Maggie, who married Oscar Karney, represented the peaceful and law-abiding element. On the other hand, Jennie, who married Bird Kelly; Lizzie, who married Robert Kelly, and on his death, John McElroy; and John W. Smothers, who married Bird and Robert Kelly's only sister, Cora, represented the discordant factions in the otherwise tranquil home."[64]

While George, Sr. might have preferred to associate with the peaceful and law-abiding elements of the Smothers clan, the Carrolls had their own share of misadventures.

George, Sr.'s and Abby's children stare at us from startlingly clear black and white photos. A young Granddad George (George, Jr.) smiles at the camera. He wears a well-manicured mustache and possesses the chiseled features of his father framed by blond hair parted on the side and cut short, and stares straight ahead with the warm, sparkling eyes of his mother, Abby. He was twenty-two in 1882, around the time the pictures were taken.

Granddad George's younger sister, Lillian, seventeen in 1882, was a beauty and hotly pursued by the Young County bachelors. Youngest sister, Nellie, was also a beauty and hotly pursued—and not just by bachelors.

Nellie is quite striking with her delicate features and blond, very curly hair cut short in front and pulled back into a French twist. Her dress is demure and comes up to her neck in loose dark folds, ruffling out into puffy shoulders over form-fitted sleeves that end in a many buttoned cuffs. The aristocratic demeanor of this lovely, young woman belies the circumstance and reality of her time and place—prettily posed in the Wild West.

In spite of her apparent youth and innocence, research on Ancestry.com revealed a family secret. According to Michael Revel, descendant of the Strahan family, Emanuel Willis Strahan Jr., a Texas state militiaman and a married man, had an affair with a teen-aged Nellie Carroll. Emanuel and Nellie's child, Theodore Lockwood Strahan (2/15/1884 – 5/28/1950), was taken to be raised by Emanuel and his wife, Mary Green Strahan. Emanuel was attacked by dogs a few years after the child's birth and died from his wounds.

Teenage marriages were not uncommon, and a time-honored solution for teen pregnancy was a shotgun wedding. "In the 1850s, 10 percent of brides were pregnant. Between 1880 and 1910, when both contraceptives and abortions were newly illegal, that figure rose to 23 percent. Although frowned upon, the pregnant bride was accepted into society. The greater concern centered on those girls who were seduced and abandoned by married men. There was no socially approved remedy for that situation."[65]

Although we can only imagine the details of this difficult situation, Nellie was certainly fortunate in that her child was adopted by his father and given legitimacy. We can only wonder at the circumstances.

The Carrolls had survived the Civil War and an inopportune pregnancy. They'd suffered great losses from which no family could soon recover. The loss of their wealth in the form of property, cattle, and slaves was nothing compared to the loss of George, Sr.'s health. Yet their losses were small by comparison. The South had been devastated. The world had changed. Slavery had ended. Hostile Indians were retreating or dead and Texas filled with homesteaders like never before.

New arrivals were a regular sight. John M. Kirby and his wife Rachael Sells Kirby arrived from Fairbury, Illinois in 1875. They homesteaded land adjoining the Carroll spread.

The Kirbys had eight children. James was the oldest, just two years younger than Granddad George and born in 1862. The second son was the tall and lanky Ira. And the oldest daughter, Mary Ella, pretty and overworked, was born in 1865 and the same age as Lillian. Younger siblings Orpha, William, Emma, Dora, and Charles Kirby kept Mary Ella much occupied as she was the oldest female child in the household and acted as assistant caregiver.[66]

Granddad George and Mary Ella Kirby might have clashed some in their youth as she was fiery and he was stubborn. But it was only a few years before Granddad George began to court Mary Ella. As we see from the example of Nellie, an attractive, young woman's innocence wasn't likely to last long in the Wild West. But George was respectful and took his time.

The Carrolls ranched in Young County until 1881 - '82 when, because of George, Sr.'s continuing health problems, the Carrolls decided to take the long journey to Wyoming through Kansas, or so the family story goes. This trip would not be like their regular cattle-drives. They planned to relocate the whole family to northern territory. Many of the Kirbys and Turks would be joining them.

George, Sr., a seasoned cattleman and rancher, in spite of the burden of poor health, and son George Jr. (Granddad George) spent much time in the saddle between Texas, Kansas and the Montana/Wyoming territory moving cattle and belongings over several seasons.

Daughters Lillian and Nellie had never been on a trail drive. In spite of the beauty and delicacy reflected in their photographs, these were girls raised by Abigail McElroy Carroll, a veteran of hard times, and they'd spent their childhood and youth in less than citified conditions. They were tough, and certainly their help was required in order for the move to succeed. They left for better climes and better times.

Lillian married Natt James in Johnson, Wyoming in 1883 at about eighteen years of age. In 1886, Nellie married Joe Dayton Thorn[*] in Johnson, Wyoming and left her past behind her. Also in 1886, Granddad George married Mary Ella Kirby in Young County, Texas.

---

[*] Dayton, Wyoming was named for Joe Dayton (JD) Thorn, who trailed cattle into the territory in 1880 and served for a time as Sheridan's banker..

SUMMER 1935

They were all crowded around the radio, even Myrtle, who had angrily denounced fighting as a waste of time if not a sin altogether, lingered within earshot. The boys had looked forward to this particular warm June evening for several weeks. This was the fight of the year.

"It's the fight of the century!" Harold exclaimed, his blond hair slick with Brylcreem and hanging defiantly over one eye. He stood with a decidedly fifteen-year-old slouch behind his father seated at the table. "It's James Braddock and Max Baer!"

Three generations of Carroll men gathered around the kitchen table. Peter sat in one of the uncomfortable wooden chairs, his cane dangled from the simple straight back. Arthur sat next to Peter, his father. Harold lurked behind him, shuffling from one foot to the other impatiently. Bob sat nearest the opened back door, enjoying the breeze. Charley Kirby sat next to Bob towering over them all, his tall, thin frame hunched into a C-shape, and still he was too tall to look anyone eye to eye. He held a toothpick between his teeth that mangled his words when he spoke so that they were practically unintelligible. Art leaned against the kitchen counter adjusting the radio's knobs to reduce static.

Myrtle thought about insisting they take their prize-fight elsewhere. She could declare this her territory and demand them out. They'd go. It was her house, after all. She told the story often. How she'd found the perfect house in the perfect neighborhood — a whitewashed three-bedroom with a white picket fence just like she'd always wanted. When the owner found out she had three boys and a dog, he told her he wouldn't rent to her.

"Well, I don't understand that. What are you saying?" she asked.

"I don't want my property destroyed," he insisted.

"Mister, would you go home with me? Come over and see my flowers!" she invited.

He took her up on the offer and after a tour of well-tended grounds, "The dog is taught not to get in the flower bed." And after showing him an immaculately maintained interior, "my boys are good boys and do not destroy property." Myrtle convinced him of their good character, and they moved to Swan Street. Myrtle and Arthur eventually bought the house.

Myrtle now found herself dawdling at some imaginary task, unable to leave the kitchen where the radio squatted, drawing her in like a magnet. As the announcer's voice became excited, the boys started shouting encouragements for Braddock, the underdog Irishman.

Against all odds, Max Baer found himself outmatched and James Braddock won the fight. Far from the arena, in Myrtles' kitchen in Sheridan, Wyoming, loud and raucous joy erupted spontaneously.

1935 was a good year for James Braddock and others of Irish descent in America who were just beginning to rise above the stigma of being Irish and assimilate into American society. 1935 was a bad year for Jews in Germany, however, as they were singled out. The Nazis established laws that created systematic segregation and hardship for those of Jewish descent and others labeled non-Aryan.

The new German laws distributed power and wealth based upon a characteristic of birth—like the old Penal Code of Ireland created by ruling England to segregate those Irish who were Catholic in the eighteenth century. The Nuremberg Race Laws were similar, if even more sinister in their intent. "The Law for the Protection of German Blood and German Honor" outlawed marriage or sexual relations between those of Jewish descent and Aryans, and "The Reich Citizenship Law" made Jews subjects of the Nazi Reich and deprived them of German citizenship. The Nuremberg Race Laws led to more than exodus and war, they were the first step in genocide of the Jews.

In Ireland centuries before, one could disinherit one's father and family and declare adherence to the protestant faith. The Nuremberg Laws resulted in exacting methods for determining just who was and who was not a Jew. It was decided that someone with three Jewish grandparents was a "full Jew" while those with less were either first degree with two Jewish grandparents or second degree with only one Jewish grandparent. Eventually, being Jewish was outlawed completely; a drop of Jewish blood could deprive someone of their basic rights as a human being. According to the Nazi propaganda, no Jewish woman could possibly produce an Aryan child.

An Aryan was defined by the Nazis as someone with blond hair and blue eyes of Germanic heritage. Based on this definition, Myrtles' mother, Mary Ella Kirby, might insist that all of her grandchildren were Aryan. Most, if not all, were born towheaded. Many had blue eyes. The Kirbys were decidedly German in heritage, and patriotically American.

The Carroll men disbanded, clearing the room and returning to Myrtle her kitchen. Peter lay down in the back room on the day sofa. Arthur and Charley sat on the porch smoking. Harold left to talk to his friends about the fight. Art sat in the living room looking at college brochures. He was determined to go to college, preferably as far away from Sheridan and the railroad as he could get. Unlike his brother Bob, he did not want to follow in his father's footsteps; he'd never work on the railroad or in the mines. He'd been very happy when they'd moved to Swan Street and away from the rail yard.

Art had been groomed for college by his grandparents, George and Ella. He was smart and capable of long hours of study. Money was a problem, but he wouldn't let it get in his way. Art was going to get an education come hell or high water. If determination was enough, he'd already have a doctorate!

Bob was determined to succeed in the military. Art was determined to succeed in college. Bob was great at sports. Art was a great dancer, and girls lined up to be his partner. A healthy amount of competition kept each of the brothers on his toes.

In July 1935, Bob completed the Citizens' Military Training Camps Blue Course of Instruction. There was a ceremony at Fort Logan, Colorado and his family attended. He was recommended for a commission in the reserve corps by Captain Smith, commanding officer. Written remarks: "Marksman Pistol score 150!"

On the entrance forms necessary to enlist in the United States Military, Bob declared his major subject in high school to be math. He participated in equitation, swimming, skiing, snowshoeing, mountain climbing, and boxing. Scouting and mountaineering had prepared him for military life. He was a natural. He listed his features on the form: H. grey eyes, blond hair, fair complexion, six feet one and one half inches in height. He was now a Private first class, but not for long; this was just a brief step on his way toward wearing officer's insignia. He was keeping alive old traditions.

Adolf Hitler was breaking old traditions and the Treaty of Versailles. The Treaty of Versailles was meant to keep Germany from again becoming a threat to the world as they had been during WWI. Hitler boldly violated it. He reasoned that mighty Britain, with an empire controlling a quarter of the world, was over-confident and not likely to declare war on Germany for violating the Treaty. Germany was a defeated nation, after all. France would never become openly aggressive against Germany on their own, not without Britain's backing. Hitler saw no need to worry about the United States. Roosevelt had made it clear there would be no U.S. involvement in European quarrels. So Hitler tripled the size of the German Army and stockpiled weapons.

STAGECOACH RULES

- *Abstinence from liquor is requested. If you must drink, share the bottle. To do otherwise makes you appear unneighborly and selfish.*
- *If ladies are present gentlemen are urged to forego smoking cigars and pipes as the odor is repugnant to the Gentle Sex. Chewing tobacco is permitted, but spit with the wind, not against it.*
- *Gentlemen must refrain from the use of rough language in the presence of Ladies and Children.*
- *Buffalo robes are provided for your comfort during cold weather. Hogging robes will not be tolerated. The offender will be made to ride with the driver.*
- *Don't snore loudly while sleeping, or use your fellow passenger's shoulder for a pillow. He, or she, may not understand. Friction may result.*
- *Firearms may be kept on your person for use in emergencies. Do not shoot at wild animals as the sound riles the horses.*
- *In the event of runaway horses, remain calm. Leaping from the coach in panic may leave you injured, at the mercy of the elements, or hostile Indians, or hungry coyotes.*
- *Forbidden topics of discussion are stagecoach robberies and Indian uprisings.*
- *Gents guilty of unchivalrous behavior toward Lady Passengers will be put off the stage. It's a long walk back. A Word to the Wise is sufficient.*

Rachel Sells Kirby

CHAPTER 12

TRAILBLAZERS

FOUND! Human bones at Angels Camp in gold mining country.

The bones indicate a male between 30 – 35 years and about five and a half feet tall. A broken thigh bone (nicely healed femoral fracture) from a gunshot wound (lead particles located by X-ray) 5 years previous to death would have made a noticeable limp...the right leg being 2 inches shorter than the left. A badly rusted H&R 5-shot .38 (produced in 1883) double-action pistol (four bullets found in chamber, one missing) was laying on the right hand.

Near the left was an empty glass whiskey bottle (pumpkin-seed flask.) Spectacles were carried for reading, and an 18K gold watch (made 1875, Elgin Watch Company #338397) with no inscription (case in perfect condition, workings rusted away, crystal shattered, solid gold chain, fob missing.) 2 suspender slides (patented 1888, Wilson Brothers, silver plated, fancy scroll pattern on front) and one suspender snap (lion and sword design) with leather remnants (C.F. Co.) and small metal suspender buckle were found near what may be heavy winter coat (herringbone pattern, brown wool.)

Pockets held one knife (multiple blades,) wooden handle of a toothbrush (boar bristle style) one bone button and pewter fly buttons (made in Paris.) Pocket money included $5 gold coin (1886 S,) $1 gold coin and half dollar (1875,) two quarters (1878 CC) and V nickel (1867.) Four short .38 rounds found in right hand pocket, with items of paper turned to mulch. No fatal trauma observed. Skeleton was found in sprawled position. ~ Nevada Online.[67]

Granddad George's uncles were among the first to drive their cattle north to Kansas markets. Census records from Gonzales County, Texas in

1870 list Lewis Carroll as a stock driver, never married. Dennis Carroll is listed as a Ranger, also never married. Columbus is listed as "beef driver, age 25."

Cattle-drives from Texas began shortly after the end of the Civil War. "In 1866, Nelson Story brought the first head of cattle through Sheridan to Montana. In 1871, six hundred thousand cattle came up the Texas trail in herds of about 2,000, each led by a wild and reckless and tough bunch of young men with great courage and fortitude."[68]

John Ira Kirby

"A cattle crew of eleven men, the trail boss, eight cowboys, a wrangler and a cook, usually managed an average herd of 2,500 cattle. Approximately two to four Black cowboys were present on most cattle drives because among them were many of the best riders and ropers in the Midwest. Ironically enough, all cowhands—whether White or Black—soon became known as cowboys which White Texans strongly resented. The eight cowboys usually rode in pairs with two in the front and rear and two on each side of the herd. Moreover, the cook was usually a retired Black cowboy and the wrangler was frequently a Black teenager who took care of the horses."[69]

The Carrolls, Turks and Kirbys took cattle to Kansas and on to Montana and Wyoming where they settled and established ranches. They also maintained their close connection with the Cockburns, and other Texas families, and returned for more cattle after their move north.

History leaves few details of the Carroll's enterprise. The official book which recorded the herds, trail bosses, etc., was destroyed in a fire in Wyoming in 1893. However, at the first reunion of the Old Trail Drivers held in 1923, a list was made from memory of the herds and their drivers by J.B. Wells, Charles Chenault, and Deputy Inspector Watts. Included on the list are "Jake and C. Carroll, four herds."[70]

Charley Kirby

A member of the Carroll's crew, Sam Grantland, "bears a singular distinction as a trail driver, in that he rode the herd one whole day, stark naked, save only his hat, and barebacked on his horse."[71]

William Kirby

Geroge Kirby

Jim Kirby

Dora Kirby

A.W. Hunter of Camp Wood, Texas describes Grantland's trail driving experiences which began in 1869:

> After the war closed, I went to work punching cattle for T.B. Malone. It was the custom of that time for cowmen to send out a hand on the range to represent him in the procedure of branding mavericks. There were no fences in those days and the country in the spring was usually full of unbranded cattle. I was paid $5 a month. The second spring another man offered me $15 a month to work for him. I turned him down, as I had made a verbal contract with Malone. In some way the matter reached Malone's ears, and he told me that he appreciated my staying with him as he needed help, and besides raising my wages, he was going to put me in charge of a shipment of cattle he was soon to send to New Orleans. He did so, and a warm friendship sprang up between myself and T. B. Malone that grew throughout intervening years. His house was my home for several years, and I was never treated better by anyone than this good man and his family.
>
> Goliad in those wild times had quite a few celebrated characters. There was Sally Skull, a celebrated female character, who often made visits to our town. She was a horse trader and her cowboys were principally Mexicans. Sally wore a couple of .45s in her belt and was said to be a crack shot. She always rode a powerful horse and spoke Spanish like a native. Who she really was and where she came from was a mystery to many. It was a current rumor that Sally was a northern woman, a member of a wealthy, cultured family, and became disappointed in a love affair, became embittered, and lit out for Texas. What finally became of her I do not know. Another famous character that frequented Goliad was the notorious Jack Helms, who it is said, had several notches decorating his gun. Helms met the usual fate of the gunman in a fight with Jim Taylor at Rancho. Bull Sutton often visited Goliad in those days. Sutton was one of the principals in the later Taylor-Sutton feud in which several men lost their lives.
>
> My trail driving experiences began in 1869. That was a long time ago, and I am pretty sure that I was among the very first trail drivers from that section of Texas. We left Gonzales County in March, starting with a herd of 1200 big steers from the mouth of Peach Creek.* Our destination was Abilene, Kansas. In the crew was Ollie Treadwell, Mercer McKinney, Frank Rollins, Ed Crawford, Gaston Cabbler, Charley Turk, McCormick, Columbus Carroll, George Reyno and myself.
>
> The herd belonged to Columbus and Jake Carroll. Our herd passed through Belton, then a very small town. We crossed the Red river at Sherman. Swam the cattle across red River. Passing through the Chickasaw and Cherokee nations we saw many tame Indians. We crossed the Arkansas River on what

---

* James Turk built a bridge over Peach Creek in 1869.

was known as the old Shawnee trail. When we reached the river, the Arkansas was up and was rising. Carroll said to us, "Boys, we have swum everything up to now, and this stream ought to be crossed, but I ain't going to tell you boys to swim this one. I can't swim a lick in the world, but we will have to go across or fall back and besides there are 20 or 30 herds behind us, and if this herd stampedes and gets mixed up with the others, we will be cutting out cattle all year."

Ed Crawford who was always ready for any emergency, fight or frolic, spoke and said, "Let's go cross, fellers. Danged if I don't always get plenty tired cutting out cattle." That steeled it. We put into the river.

I was on the right hand point and Ed was pointing the herd on the left. Before starting the herd in, we dismounted, pulled off all our clothes, took our saddles off our horses, and with nothing but our hats on mounted our horses barebacked. We got the herd across with little trouble. The boys that were supposed to be following on my side trailing the herd across the swollen stream got cold feet and did not put into the river, and I had to swim my horse back and forth until the bulk of the herds got across. Finally, my horse gave out, and I had to take to the water. I swam to the shore and got out by grabbing a willow limb. About 300 yards below there was an island in the river on which about 500 steers landed, along with my horse. After resting a few minutes I swam to the island, got on my horse, shoved the cattle off, swam them across and joined the main herd. Along with Crawford. This occurrence was in July. The boys on the other side of the river were building a raft on which to cross, and which besides its cargo of human freight, also carried our bedrolls, provisions, and most needed of all just then, mine and Crawford's duds.

We got the herd across all O.K., but there we were with no more clothes on than Adam had in the Garden of Eden. Ed mentioned this and wound up saying that even at that we had Adam bested, because we were hossback.

Well, Adam had shade trees and fig bushes to protect him from a July sun – and we had nothing but our hats. And besides that herd had to be held at any cost. All day long we rode, and each minute the sun got hotter and the blisters grew bigger. During our herd riding we discovered a plum thicket loaded with ripe luscious, juicy plums. We stayed in the plum thicket for an hour or so, while our herd was grazing and headed in the right direction. Leaving the plums, we struck out to point the herd, and in reaching the top of a hill and looking down in the valley, Ed discovered a party of about 500 Indians. Ed suggested that we retreat to the brush. I said "No. Let's stay put." The Indians came up and one of them shook hands with Ed, and tried to shake hands with me but my horse being skittish he had to give it up. They rode into the herd and shot down 12 steers, skinned them and packed them off.

Along about 6 o'clock in the evening the raft came across and Ed and I got some relief. We needed relief. We were sunburned, blistered, saddle sore, skinned and briar scratched. The next morning we started the herd on its way, and had traveled only a few miles when we met another bunch of Indians. The chief of this bunch wanted a tax for crossing the herd through his country. Carroll dissented and refused the request, informing him that he had donated 12 steers already. This chief spoke excellent English and informed Carroll that their wives and children were starving and almost destitute and must have fresh meat. Carroll after consulting with owners of other herds that were behind us, consented to give the Indians some beeves, provided the Indians would keep out of the herds. This was agreed to and the herd owners sent men into the herds and cut out steers that were sore-footed and road-foundered, and turned them over to the Indians, who were very well satisfied. This ended our trouble with Indians.

We reached the Smoky River and found it bank-full. Securing skiffs, we embarked in them, swimming our horses along side. The herd had already been started across. Arrived at Abilene without any mishap, where Carroll sold the steers at $15. Us trail drivers got $1 a day and board, with our horses furnished us. Abilene was a small burg then. The population was about 300. The town had three or four saloons, and two or three stores. Whiskey was 5 cents a drink. Some of the boys saw their first railroad at Abilene. And some saw their first sod houses there.

This covers my trail driving experience. I was among the first to go up the trail, and at times it was a ticklish job. No blazed trails — and Indians ever so often. We had the old North star, youthful spirits, and even with the dangers and other hazards, we had worlds of fun all the way from Goliad to Kansas.[72]

It was rough, hungry work and cowboys grew accustomed to corn meal biscuits and salted bacon. They fashioned saddles from whatever was at hand, pieced together chaps from scraps, and braided rope from the hair off of their horses. They had little or no protection from the elements and stretched out under the stars to sleep, their Colt sometimes the only companion at their side.

The cowboys crossed into Indian Territory on these cattle drives and sometimes encountered hostility. The Comanche had wielded great power in Texas, resisting the unrelenting incursion of white settlers. The tribe was isolated mostly in west Texas now, but they were hungry. Their warriors now acted out of desperation, with vengeance.

After a raid in 1870, Kiowa and Comanche warriors surrounded Lewis and Jake, Jr. and their Texas cattle herd as it was bound for the Cheyenne and Arapaho Agency at Camp Supply in Oklahoma. The cattle were stampeded and hundreds killed. Both Carroll boys died giving their lives in battle, and are buried in the family cemetery south of Gonzales, the year of their death carved in stone; stone that was once rough but is now smooth to the touch and fading with time, as is the memory of Lewis & Jake, Jr. and of cowboys and their stories.

And just a few years later in 1876, the Sioux and Cheyenne Indians led by Chief Sitting Bull & Chief Crazy Horse defeated Lieutenant Colonel George A. Custer and his Seventh Cavalry of over 260 men in the Battle of the Little Bighorn in Montana territory.

Yet most of the Indians the cowboys encountered on their cattle drives were hungry, destitute, and defeated by the U.S. government policies. The cowboys referred to them as "tame." Of greater risk were rustlers and outlaws.

In September 1881, George Granville, a young man of 21, rode in front of the procession of wagons carrying his family and belongings northward to Kansas. They were destined for Wyoming hoping the cooler, dryer climate would aid his father's health. This was not a cattle-drive, although they were taking some beeves to market in Kansas, and a few on to breed.

They loaded up all their belongings in wagons and buggies, packed the mules and bridled the horses. An arduous journey of many months lay ahead of them.

They left with the clothes on their backs, as the family legend went. It must be noted, however, that they possessed many horses, mules, wagons, and buggies. They were certainly not destitute like many of the Indians they encountered. But these were dangerous times and every place was distant. Young George Granville had to earn money, and whenever a stop was made he hustled for a job and did whatever he could to bring in a dollar.

But it wasn't just men on this adventure. In addition to George, Sr. and his sons, George Granville, Solon and Fred; Mother Abby, and oldest daughter, Lillian, traveled with them. The women gathered wood, cooked and sewed and provided medical care and advice, whether welcome or not, as well as coffee and solace on the long and dusty road.

They wintered in Kansas, staying with family near Paola.* George found work, "to keep them breaded." On their way to Cheyenne, they passed through Indian Territory where they encountered "difficulties" with cattle rustlers. George, Sr. had learned the habits of rustlers when he trailed cattle through Louisiana and Texas to Abilene where the Kansas Pacific Railroad shipped the cattle east. His knowledge and use of "secret hobbles" on mule teams and horses helped overcome problem encounters with rustlers.[73]

Other brave travelers followed the same route. After they took their

---

* Exactly who they stayed with is not recorded. The Kirbys may have had relations in Kansas. John M. Kirby, Ella's brother, lived at Laclede Hotel in Paola, Kansas and died in 1900 of a stomach disorder, possibly cancer, according to his obituary.

leave of Cheyenne heading north, they overtook Mrs. Nettie B. Held, traveling a bit more slowly with her mother and small daughter called Birdie. Further on, they caught up with Nathaniel Huntington and his family. They caravanned together the remainder of the way and all three families arrived on June 18, 1882 in what is now Sheridan, Wyoming. They camped that first night on what is now known as Marion Street.

Sheridan Rodeo, 1948

SUMMER 1936

Rodeo! A new tradition with an old history, rodeos brought out the whole town. Granddad George and Ella, Arthur and Myrtle, and Art and Harold all filed onto the crowded bleachers, sitting close together on the smooth wooden planks. Excitement filled the air. It was time for the fourth annual Sheridan Rodeo. Everyone turned out for the event. Even the Crow were represented, blessing the fairgrounds with their peace pipe ceremony.

Granddad George and Grandma Ella had taken the boys to the first ever Sheridan Rodeo in 1931, a rag-tag affair put together with contributions from local shopkeepers. In 1932 Crow and Cheyenne representatives smoked a peace pipe as part of the night festivities. In 1933 there'd been no money for a night show, but the rodeo went on in spite of hard times. And this year, 1936, with donations of money and the hard work of volunteers, a new grandstand was added and other improvements were made. It was an event defiant of the depression. And to top it off, Sheridan had elected the first Rodeo Queen! The crowd would stay for the crowning.

People milled around between events. "I'm off to get lemonade," Myrtle announced, "Would you like anything, mother?" she offered Ella. Granddad George had already left to look at the livestock; Art, and Harold had taken advantage of a good opportunity for girl watching, and loped off to inspect their attributes.

Art was an excellent horseman but preferred to be a spectator at rodeos. He'd seen quite a few rodeos by the time he was eighteen and could clearly remember his first like it was yesterday, a rodeo in Edgemont, South Dakota that he attended when he was only nine years old. The bucking broncos and the bull riding events had kept him riveted.

They gathered back in the stands for the next event. Ella's small frame boasted a big voice as she shouted her encouragements to the bronco buster stepping up to the coral—one of her grandchildren, Granville's and Alta's child—getting a last minute "go get 'em!" from his cousins Art & Harold, before leaving the gate.

Ella was a tiny but tough pioneer woman who was sweet and mannerly, even when you made her mad, but that was not recommended. She used to be young and spry and so desirable a man would ride horseback half a continent to ask for her hand in marriage. Ella sometimes told stories about Granddad George within his earshot which he deliberately ignored and usually "harrumphed!" But when their eyes met, you could see the sparks. For Granddad George, Ella was still his young bride, generous, kind and capable—and forever twenty.

When George tipped his hat to her that windy day in 1881, he knew he would return to Texas for Mary Ella Kirby.

"Pa! Let me do that!" George, Jr. hollered. George, Sr. was hefting baskets and trunks into the wagon and arranging them in his meticulous manner.

"We can't get much more on here," George, Sr. yelled back, ignoring his son's plea and lifting the heavy loads as if he were a young and healthy man of twenty and not a wounded civil war survivor in his middle forties. The best doctors around, including his brother-in-law, Dr. Johnson, had advised him that moving north would help his breathing. He just couldn't catch his breath anymore. And of course the bullet was still in there, taunting him. It couldn't be removed without costing him his life, and leaving it in seemed to be costing him, at least, his quality of life. What irony. He hefted another basket of kitchen ware onto the platform. George, Jr. was beside the wagon now and handed up the last of it.

"Ladies! Too much," George, Sr. protested as they brought still more packages out onto the porch. "The wagon is full."

Abby and Lillian, mother and daughter, looked at each other knowingly. There was some way to bring these packages along. They'd figure something out.

Solon brought the horses already bridled. George, Jr., Lillian, Solon, and Fred and parents Abby and George, Sr. were traveling to new vistas again.

"Remember George and I will be back. Leave what you have to. The Kirbys will watch over the place." George, Sr. assured Abby, knowing they would also keep watch over daughter Nellie who currently was a ward in the home of Mr. and Mrs. Strahan.

It had been a number of years since George, Sr. had been on a cattle drive, but he knew the ropes. They brought a small number of cattle with them this trip, along with horses and their loaded-down mules, one carrying a couple of extra packages George hadn't noticed when he secured the animals.

Abby had the reigns of the wagon. George, Sr. took the lead on his 15 hand steed. Solon and Fred kept eye on the loaded-down ponies, and mules attached to each other and the wagon. Lillian rode her palomino next to her older brother, George Jr., the two of them taking up the rear of the procession. They also had a crew of Kirby cowboys with them, driving their own small herd north. The plan was to sell a few head of cattle and winter in Paola, Kansas. George G. would work to earn money for the family's travel expenses in the spring. John, Charles, and Ira Kirby were already a half day ahead of them; Ira was to accompany them all the way to Wyoming and Charley planned to return to Texas.

The roads and byways from eastern Texas up through Kansas were now pretty well secured from hostiles and rustlers. Although the men were well prepared for any trouble, they traveled disturbed by nothing more than bad weather and vermin. They camped at night and George Jr. set up a canvas tent for the ladies. The men slept under the stars, and when it rained, under the wagon. George Jr.,

Sheridan Rodeo, year unknown

Solon, and Fred took turns guarding the camp, although Solon usually fell asleep, causing an exchange of words as his older brother, George, took him to task for it.

Many tales were told around the campfire as they settled their bones for the long nights. George, Sr. told tales of riding the trail when he was young—tales of cattle rustlers and Indian attacks. George, Sr. had learned the habits of rustlers when he trailed cattle in his younger days. He told of their tricks. He also could identify all of the animal sounds in the woods.

The Carrolls made it safely to Paola with horses and stock in tow. The harsh Kansas winter with relentless snow and blowing winds helped prepare them for even rougher Wyoming weather to come. They boarded with the Kirby cousins who made space for them in "the old place," essentially a shack out back that used to be their home before they built the new one. Used mostly for storage and guests, it included an occasional raccoon that might soon be in the dinner pot. It cleaned up pretty well and a wood fire kept it cozy. They sold most of their stock and George, Jr. took a job as a hand in the local livery stable. After wintering fairly comfortably, they left Kansas and headed west in the spring of 1882.

One moonless night on the Overland Trail, while Solon slept peacefully under the stars with his rough woolen blanket pulled up to his ears, he was suddenly awakened by a nudge on his shoulder, then a harder one on his shin.

"Ow! What is it?" he sat up awake.

Fred was pointing over the dying embers of last night's fire and beyond, into the darkened brush. "There's something out there."

"Oh, you fell asleep and were spooked by an owl!" Solon began to lie back down and ignore him when he heard it, too. "Something's spooked the horses!"

Both boys looked at each other, thinking the same thing—Rustlers! They turned to go get Pa, but he was standing behind them, shotgun in hand. "Shhh," he warned. George, Jr. appeared now, looking ready. George, Sr., using hand signals, gave the boys directions to surround their meager, make-shift coral.

The night was dark, the moon only a sliver in a cloudy sky. Fred inched his way through the trees, gun in hand. He was a good shot and knew it, but he was also half asleep and scared. Solon moved brazenly forward on the other side of the pen, trying to keep up with his older brother, George Jr., who'd disappeared into the inky night. They heard shots, and everyone ran to the sound and found George Jr. with smoking gun in hand, pointing into the brush.

"They ran that way!" George G. said between breaths. He'd run after them a fair distance before he'd fired a warning shot.

Solon darted off in the direction he was pointing only to be stopped by George, Sr., "No. Don't waste your time, son. I've got a better idea," he said, and they headed back to camp.

George, Sr. had knowledge of "secret hobbles" for use on mule teams and horses. Hobbles keep the animal from moving too far, sometimes from being able to move at all. There was the very restrictive knee hobble, which doesn't allow even a small step forward. Spider hobbles strap all four legs together and allow the horse to stand in a normal position, but they can cause serious injury if the animal fights them, unless they have an efficient quick-release mechanism. George selected a variant of the practical serving hobble. He tied a strap around each hock and connecting these by a cord running through a pulley on the horses' neck collar. This allowed forward movement for the animal, and were essentially invisible to any fool rustler. Even if someone tried to run off with the animals, they'd soon discover the animals craftily ensnared. George and the boys could reach the coral before any rustler might free the hobbled animals. Morning arrived without further incident, and they hit the trail again.

They arrived in Sheridan unmolested and began building their future in the Bighorn Mountains. George, Jr. returned to Young county, Texas the following spring and drove a herd of cattle north. He also paid his respects to Mary Ella Kirby, as he was traveling in the company of her brothers, George Ira and Charley Kirby. George Jr. stayed several nights on the Kirby ranch allowing him many occasions to make eyes at Ella.

On Feb. 15, 1886, five years after leaving Young county, a strapping and confident twenty-six year old George G. Carroll returned to Texas and married the lovely Mary Ella Kirby. A large crowd filled the Baptist church and there was a fine party following, and then George, Jr. and Ella left together for that long journey to Sheridan, Wyoming. They homesteaded near the abundant Rosebud River in a thick forest. Fat cattle grazed peacefully nearby.*

---

* George and Ella later lived in Dayton and Sheridan, Wyoming where they were part owners of the Sheridan Manufacture Co.

Mary Ella Kirby Carroll

George Granville Carroll

CHAPTER 13

CODE OF THE COW COUNTRY

In the 19th century U.S. West, the cattlemen, sheepmen, and farmers all despised each other. Similarly, throughout human history farmers have tended to despise hunter-gatherers as primitive, hunter-gatherers have despised farmers as ignorant, and herders have despised both.
~ Jared Diamond, Guns, Germs and Steel.

Granddad George and Mary Ella didn't move to the wilderness all alone. As with earlier family migrations, they immigrated north as a clan. Ella's older brother, John Kirby, accompanied the Carrolls up the trail. A few years later, in 1889, he moved to Sheridan with a small herd of cattle and settled with his family on the Rosebud Creek in Montana where the Kirby Post Office is today.

The Turk brothers also located in the area. The thriving city of Dayton was the location of the stage coach station and was the way station to the West, sitting right at the foot of the Bighorn Mountains. George & Ella settled comfortably in the aspen forest along the Tongue River. George, Sr.'s sister, Nellie, left her past behind her and married J.D. Thorn. Thorn purchased a 400 acre ranch above Dayton, Wyoming.

The Carrolls, Turks, and Kirbys had not left feuding and hell-raising behind in Texas for peace in an idyllic Shangri-La. In 1889 Butch Cassidy and the McCartys robbed another bank, making off with over ten thousand dollars. As usual, the gang never fired a shot but lawmen formed an angry posse causing Cassidy and the others to go into hiding. A favorite place to disappear was Wyoming. Cassidy eventually decided to follow the straight and narrow path and work for a living. He took several jobs on ranches working as a cowboy. When he got a job as a butcher in Rock Springs, Wyoming, he was given the name "Butch" Cassidy.

Ella Kirby Carroll was no stranger to outlaws and hostiles. In addition to homesteading on the Brazos as a child, she was the daughter of Rachel Sells Kirby who was born in Dixon, Illinois in 1839. Rachel's parents were killed by Indians. Passed down from eldest daughter to eldest daughter in this family is a counterpane—a decorative piece of cloth for the mantle. It has survived harsh travel conditions, hostiles, outlaws and time. [74]

Myrtle, Jewell, Orpha, & Ednis Carroll

Ednis, Jewell, Myrtle, & Orpha

Granddad George and Ella Carroll raised their five children in the shadow of the Bighorn Mountains. Granville, the eldest son, was named after his father [and the Earl of Granville]. Second born was Myrtle—Bob's, Art's & Harold's mother; who helped to raise her younger sisters: Jewell, Ednis, and Orpha. Granville worked alongside his father bringing in the herds and cattle ranching. All the children hiked the steep mountain trails, fished and hunted in the fragrant meadows, and swam in the icy waters near Dayton. But they dressed in their best for the Sunday sermon.

A 1911 snapshot shows Granddad George's and Ella's four girls posing and smiling in the General Store in Dayton with its high, painted ceilings and glass cases filled with jewelry and semiprecious stones. It is Christmas time, as noted from the colorful, handmade decorations strung around the room.

Outdoor people, most Carroll family photos show folks posed outside, often dwarfed by the magnificent landscape. A 1929 snapshot shows Jewell and Myrtle camping in the high country up the Dayton-Kain road at Woodrock. They wear calf-length skirts, boots, and fashionable hats, and collect sticks along the river's edge. Another photo shows Myrtle holding animal skin blankets near two wigwams at Woodrock.*

An often-handled photo shows a young Myrtle and Arthur beside a stream. He wears striped overalls and top hat. She wears denim, work boots and an old, bent-brimmed hat. He's leaning back on a large boulder while she lies back between his outstretched legs. They're wrestling, or he's tickling her, and they're both laughing.

Arthur and Myrtle were married on August 18, 1914. Myrtle's wedding photo shows a lovely twenty-one-year-old in white silk brocade and lace. She wears a round, gold locket which contains the photos of her grandparents, George, Sr. and Abby.

Myrtle's older brother, Granville, was married just a few years later. At twenty-eight he met and fell in love with smoky-eyed Alta, a school teacher from Colorado. They married in Spokane, Washington in 1918. No East-coast prissy, Alta was raised in a rough, natural environment like Myrtle, and so might be able to enjoy life on the range with a cowboy.

Granville was a cowboy in the traditional sense. He worked as the foreman on the extensive properties owned by the Coffee Brothers in Washington and Oregon. And he was raised by Granddad George to be an expert on the range.

Dee Brown, in The American West, explains the role of the foreman or range boss:

"The success of a roundup usually rested largely upon the shoulders of the range boss, who in the early days was the ranch owner. As big ranching developed,

Carroll girls camping up the Dayton Kain at Woodrock, 1928

---

* The Dayton Kane road connected the mountain cities Dayton and Kane. Kane boasted the only ferry crossing of the Bighorn river above Greybull, Wyoming. It had a bank, hotels, dance hall and saloons to accommodate the many cowboys who came to barter cattle. It flourished into the 1930's but was flooded in 1967 and now lies under the Bighorn lake.

the owner would select an experienced and respected cowhand for the job. During a roundup, the authority of the range boss was as ironclad as that of a ship's captain, but to keep the job he had to know how to manage three of the most unpredictable members of the animal kingdom—cattle, horses, and cowboys."[75]

Granville and Alta face the camera on their wedding day. She glares daringly at the camera. Her thick, black hair is pulled back and her eyes sparkle. Granville faces the camera confidently. His neck is restrained and held high by the fine white collar he was certainly unused to and uncomfortable wearing. A fresh haircut, cropped too short, reveals his youth.

Alta wrote about her experiences as a young bride in the unbridled west. She was born in a sod-roofed cabin on her family's homestead near Merino, Colorado. Her mother claimed the sod roof was "very nice." She said when it rained it was always very dry inside the house, but the next day when the sun began to shine it rained inside. "You could go out when it rained."

Granville's grandfather, George, Sr., survived thirteen years in the Montana/Wyoming territory. He died at age sixty-three in Sheridan on Nov. 14, 1895. His wife, Abby, lived another thirty years and continued to offer consolation, her sharp wit, and advice to Bob, and others, until her death in October, 1925. According to her obituary, she "just lay down and died" that Thursday afternoon.

Myrtle & Arthur Carroll

CODE OF THE COW COUNTRY
S. Omar Barker

It don't take such a lot o' laws
To keep the rangeland straight,
Nor books to write 'em in,
'Cause there are only 6 or 8.

The 1st one is the welcome sign,
Written deep in western hearts,
My camp is yours and yours is mine,
In all cow country parts.

Treat with respect all womankind,
Same as you would your sister,
Care for neighbors strays you find
And don't call cowboys "mister".

Shut the pasture gate when passin' thru
And taking all in all,
Be just as rough as pleases you,
But never mean or small.

Talk straight, shoot straight,
Never break your word to man or boss.
Plumb always kill a rattlesnake,
Don't ride a sore-back hoss.

It don't take law nor pedigree to live as best you can.
These few is all it takes to be a cowboy and a man.[76]

FALL 1936

In December 1936, after he was appointed Second Lieutenant of Infantry, Bob came home for Christmas. The old photo albums were dusted off and new pictures added. The new right next to the old.

They stand together, solemn of countenance, respectful of a serious occasion. At least, most are respectful. Granddad George has clearly said something that made the finely dressed ladies cast their eyes downward and display emotions ranging from amusement to shock, their expressions frozen in time by the miracle of photography. They posed for the camera wearing the latest fashion of the day, including hats. Especially hats.

They are lined up in two rows. Eleven stand behind and four in front; one man sits. They all wear early 20th century garb—men in vests, ties, and pocket watches; women in calf length dresses and lace-up boots, gloved and carrying handbags of varying sorts. On either end stand tall men: Granddad George and Ira Kirby both facing slightly in, framing the group. George and Ira sport fedoras. Charles Kirby stands in the back row center holding an umbrella, also wearing a fedora. Ella Kirby Carroll stands between her husband and brother mostly obscured behind a large hat worn by another woman.

The photograph is well proportioned, everyone accidentally or manipulatively placed to create symmetry. The parasol is centrally placed (for shade, if only for a few people, from the unforgiving sunlight), and there are a wide variety of hats on display. The hats add color to this black and white photo, and attitude.

There are a few people conspicuously hatless: second from the right, the lovely Myrtle, all in white with a wreath of wildflowers demurely placed in a halo shape on the dome of her head; and Arthur, the only one seated, nicely dressed in suit and tie—his Sunday best—on the grass holding his knees. Three others, a man and two women, stand in the back row shamelessly bareheaded.

Everybody else wears a hat — big hats, little hats, gaudy hats, simple hats, hats with flowers, hats with feathers, hats with ribbons, hats with polka dots! One woman wears a hat that looks like an upside down basket made of silk and tied with a gigantic bow. Eleven hats on display, smiles below, the proud members of the Carroll and Kirby clans swagger and sashay on Myrtle's and Arthur's wedding day.

The old photo is placed carefully in the album alongside another. More people in hats, pictured this time on a horse-drawn wagon. The wagon is uncovered and very long with spoke wheels and a railing of sorts. It holds thirteen people who stand behind and four others who sit with their legs dangling from the platform. Ira Kirby stands to one side of the wagon, the only one on the ground. Charles stands in the back row, this time without any parasol and his sister Ella again at his side, and men stand on both ends of the group, for added symmetry, of course. This is a reunion of the proud Kirbys.

Hats

Carroll Kirby clan reunion

Alta Carroll

CHAPTER 14

JUST MARRIED
Written by Alta Carroll [77]

Granville asked what was wrong. He answered, "I lay down to take a drink out of the crick. After I was full I turned my head. There right at my side, drinking with me was a rattler. I got out of there quick. Then I went back and killed him." He pulled a rattle out of his pocket. It had eight rattles and a button!
~ Alta Carroll

"WHERE AM I?" I WONDERED. AS I AWAKENED AT DAYLIGHT THAT DAY IN JUNE, LOOKING AROUND I SAW THREE log walls with a casement window at the head of my bed. "What is that?" A huge tarp divided my room from the rest of the cabin. "Now I remember, this is Mountain Home."

I had married Granville Carroll a few days before. We had spent several days in Yakima visiting with Granville's employers and many of his friends.

We had been married at my home in Spokane just seven months after we met the day the Armistice was signed, November eleventh 1918. He and a small crew of men were moving a herd of cattle from the summer pasture to the feed grounds where I was boarding. They stopped by the school house where I was teaching, to bring me news of the Armistice.

Stanley Coffin, one of my husband's employers had told us that if either of us could drive, there was a Ford at Palmers Barn in Ellensberg that we could use. Granville said, "I can't drive." Quickly I said, "I can." I had driven all of ten miles but I knew we needed that car.

When we got to Ellensberg we went to Palmers. The car was a 1917 model T Ford coupe, converted to a truck. Spark and accelerator operated by hand as did the crank for starting. It had been left there for an overhaul so we thought it was in good condition.

We headed toward Mountain Home after we had filled the car with gas, through the small way station of Kilatas, then north east winding through the flats to the foothills. The elevation here was about 3000 feet. Then we started twisting and turning upward fast toward the summit of Klockum Pass. The elevation at the summit is 5280 feet, traveling distance of five miles. At the foot of the mountain was a small house. Granville said this was the last house we would pass until we came to Mountain Home.

After about a mile of climbing the car stopped. Neither of us knew a thing about repairing it. So after cranking it until we were sure it would not start again we just sat and discussed it.

"Isn't that a car I hear? Yes I think it is." Another model T drew up and stopped. A man and woman were in the car.

"Need some help?" Of course we did. He cranked it and said, "I heard you pass our house and told my wife you would never make it. I think it sounded like the timer. Do you have an extra one?"

We told him the car had just been overhauled and we did not know we might need one. He took an extra one from his car and installed it then watched us start it. He would take no pay but the cost of the time. He followed us to the next turnout then went on ahead, assuring us that we would be all right now and that he was "in a little of a hurry." (Later that summer his wife told me they had heard that the Wenatchee sheriff was headed toward their still and they were on their way to move it across the county line. They made it and we were glad.)

We climbed for about five miles by speedometer and 2000 feet up when the road leveled off and the sun went down. We were on a narrow road in deep timber. The headlights were on the magneto, the tail lights on kerosene, and we lighted them with a match. As I did not know the road I had to go very slowly. The slower I went, the dimmer my lights. Granville stood up and would say, "Turn right", "No left." "Right", "By Gosh! I can't even see the road." I stopped at Yakima. The temperature had been 110 degrees when we left. Snow was not far away from us now and the night was getting cool. Our baggage had not arrived on the train, so we were very thinly dressed for spending the night. Finally Granville said, "I think I could run ahead and you follow me." He gave me his coat and

Glory Be, his shirt was white. I followed as slowly as possible and after an eternity we came to a park. Granville stopped. "Here is the gate of Mountain Home. We will leave the car inside the gate and walk in."

Just then we heard a "Yippee," then a whole chorus of them, so with the escort of six cowboys we drove in to the cabin, "Mountain Home," how beautiful it still sounds and how wonderful the memories of the people who lived nearby (which is a radius of about forty miles more or less.)

A fine supper was waiting for us. I renewed my acquaintance with Poke Smith and Lloyd Cochrane who had accompanied Granville to the winter feed ground near Beverly.

The men insisted on doing the dishes while we explored the cabin. It was about eighty feet long and twenty five feet wide with eight built in double bunks, four on each side making sixteen double beds, and a huge fire place opposite the front door, then our bedroom which I have already described. At the other end of the cabin were cupboards, a stove, long table and rustic chairs and benches. Coleman lanterns and one lamp furnished the light.

Now that I was fully awake I arose. Breakfast was already made and I stepped out on the back porch to wash. We were in a large, breathtaking mountain park surrounded with beautiful evergreen timber. About twenty feet away was a big snowdrift. I was exhilarated with the beauty, also the coldness of the wash water. Spring water was piped to the edge of the porch then ran away in a beautiful mountain hill stream, sparkling and clear. Mountain flowers broke the greenness of the park. What a gorgeous early dawn.

The call, "Come and get it or I'll throw it out," broke the stillness and I came out of my daydream, realizing how hungry I was.

Following that hearty breakfast, Granville spent a short time outlining the work with the men. There were about ten men including the fencing crew of four whose names I cannot remember at this time. The cowboys were Poke Smith, Floyd Cochran, Eddie Ingersol, Charlie Ross, Len Rickman, and Bud Oversby, a fourteen year old orphan boy who was the camp mascot.

Poke Smith's first remark to me when I came in had been, "Alta did you know that when I left that school house at Beverly I told Granny [Granville] that if he did not bring you to Mountain Home he was a bigger fool than I thought he was?"

I said "Oh! Did you?" I did not tell either him or Granville that I heard it and so did my pupils.

When Granville returned he showed me my outfit, saddle and saddle blanket and bridle and spurs. He then introduced me to my personal horse, Casey Jones. He was small, about 900 pounds, very gentle, easy gated, and very fast. During our courtship Granville taught me to properly saddle, bridle and mount a horse. This was very necessary in case I should be alone and need to ride, so he insisted that I do so now in order that Casey became acquainted with me.

Granville had full charge of the cattle outfit which belonged to the Coffee Brothers at Yakima, Washington. This consisted of about 100,000 acres of pasture and timber land and at this time he was running 2000 herd. Sheep were pastured on the highest pasture.

We rode to a high point overlooking much of the land. Granville showed me that the streams all crossed the road near the cabin. If I remembered that, I could always find my way home. "Also give your horse his head and he will bring you home if it is dark or stormy," he said. He showed the land marks that could be seen from various places. We returned about noon. We had ridden about fifteen miles.

After dinner we rode out to inspect the fencing work and found the men were working on fence through shale. He helped them to solve some problems there. On the way back to the cabin, my horse gave a snort and started to bolt. After gaining control, on higher ground and around a bend in the trail we saw the first bear I had ever seen in its natural habitat. A horse can smell a bear a long way off. The bear saw us and was as frightened as our horse had been. It ran down hill and soon disappeared from sight. Bears run in rolling gait that looks awkward, but one can outrun the fastest horse.

Granville explained that Casey would never let another horse pass him when he was running. Also that no one other than himself was to ride him. He would need him on a few occasions such as when they were gathering the range horses. He said, "How I came to love that little strawberry Roan!" Early the next morning we left in the Ford for the Horn Ranch. This was the lower ranch about fifteen miles below by horse trail but much farther by road which followed the Kloockum Creek. It was even steeper than on the Ellensberg side, but the terrain was not so rough. It was an inspiring sight and I was glad we had to go slowly enough that I could enjoy the beauty. It took about two hours to make the trip down to the Columbia River. Looking across the river I saw a man pushing a raft with a pole. "What is he doing on that raft?" I asked. Granville laughed. "We will soon be crossing the river on that raft," he said.

As he neared the shore I noticed it was much larger than I thought. The operator said he could carry three fords or two larger cars on it. We drove on and stopped in the center as directed. The ferry started to turn as he poled it out into the landing. I thought for a moment we had broken loose from the cable but I was wrong. The raft was heading into the current and slowly turning downstream controlled by the cable and a system of pulleys.

When we were about half-way across the operator took a dipper with about a ten foot handle dipped into the water. He brought it full of sparkling cold water. The temperature was about 110 in the shade but, no shade. No drink ever tasted better.

As we came to the approach we drove off. Two cars were waiting to cross. We drove to the small town of Columbia River Siding. We picked up our mail, groceries and a trunk of mine that had been sent from Spokane. Recrossing the Columbia we started down river toward the Horn Ranch, so named from the shape of a cliff near the ranch. We crossed three streams that had broad canyons. The road went down and through the canyons and across a small bridge then back up the other side, making U shaped curves. Each canyon was about twenty-five feet deep. As we came up out of the first one our car slowed then barely made it to the top where it stopped and would not move but the engine kept running.

The road continued to run very close to the Columbia and we decided to walk the rest of the way to the ranch. A very narrow road had been built with three switchbacks across a mountain of shale. We followed the steeper horse trail that eliminated the switchbacks. My every day shoes were in the trunk and the ones I had on had Louis Fifteenth heels. I was constantly stopping to pick out small rocks. I've hated similar types of heels from that day.

Finally we looked down into a valley to the Horn Ranch. There was a huge Spanish style house with red tile roof and grey stuccoed sides. A matching garage was nearby. Large corals and a bunk house sat beyond a garden, orchard and grapevines.

There were no phones in that part of the country but people have a way of overcoming problems. After we reached the ranch Granville went down to the river which was about a quarter mile wide at that point. He "hallooed" three times. In a moment he was answered from the other side. The neighbor was a mechanic who rowed across. There were horses in the coral so he and Granville rode up

Granville Carroll, about 1918

to the car. A quick inspection soon showed that some washers that had loosened on the axel had dropped down and worn it in two. Hershel did not have an axel on hand so he ordered one from Wenatchee.

In the middle of that night I suddenly awakened. I had dreamed of that left axel breaking just before we reached the top of that dip and we were falling back down the canyon. I awakened and thanked goodness we never hit bottom!

The next morning Hershel and Granville pulled the car down to the Horn for repairs. In the meantime the cowboys had arrived from Mountain Home ready to make the final ride of the draws to throw the cattle they had missed in former rides into the summer pasture. Two men were assigned to each draw. Floyd took one side and Granville and I the other. We finished our side first and sat down in the shade to await Floyd.

A very pale Floyd arrived soon. Granville asked what was wrong. He said, "I lay down to take a drink out of the crick. After I was full I turned my head. There right at my side, drinking with me was a rattler. I got out of there quick. Then I went back and killed him." He pulled a rattle out of his pocket. It had eight rattles and a button!

We ate a lunch of crackers, Vienna sausages and sardines and the coldest spring water imaginable, then took the cattle on up the canyon putting them through the pasture gate by Mountain Home. By that time the others began coming with their cattle and Granville counted them through the gate.

Two more days work of this kind completed the job. By then the car was repaired. I drove back over a road I would never have attempted in other circumstances. It was full of switchbacks, narrow and steep. It had never been used for any vehicle except wagons. There were huge logs at each of the curves. After what seemed ages we reached solid ground. For the first time I stopped and looked down at the Columbia, it was about one hundred feet below me. I lay my head on the steering wheel and shook! Then I decided I was now an experienced driver and drove on.

Arriving at Mountain Home things settled into routine, as much as they ever became routine. I took over the cooking and house keeping. I had my first dinner guests, some Indians who were hunting edible roots and plants. They ate dinner with us and in exchange left me some camas roots which are bulbs about one inch in diameter. They look and taste very much like small Irish potatoes although they are much sweeter. After that many hungry people stopped by for food or shelter, which was never denied anyone. Several days I have prepared five extra meals for people who were stranded on the road. I soon learned to cook in quantities, staples that could be warmed up quickly.

The crew gave us a dance in our honor on Saturday evening after our arrival following that roundup. About seventy five families along the Kloockum and surrounding ranches came, mostly by buckboard, buggies, horseback and a few cars. Granville allowed the men to take the car to bring the musicians and their instruments. They arrived with an organ, violin, banjo, mandolin and guitar.

Granville and Poke Smith stood at the two doors and took all six shooters and moonshine and locked them in a chest as a precaution against rowdyism. Dance tunes such as Smiles, Rainbow, Redwing, Tipperary, Smile the While and various waltzes, fox trots, and two steps were played. At intermission Granville was asked to play the mouth harp. The banjo player enjoyed it so much he accompanied him. Then one or two of the men started to jig and Granville joined in as he played. This was unplanned and really brought down the house.

Shortly after dancing was resumed the Coffin Brothers Sheep foreman yelled "Powder River Let-er-suck!" He started shooting. He had cached both six shooter and liquor outside and had slipped out while no one was watching. One woman jumped on a chair and yelled, "Damn you, stop!" Men soon subdued him, but the dance was over. Refreshments were served and the guests prepared to leave.

Then Bud, the horse wrangler, came in to say that the gate had been opened and many of the horses had escaped to the big horse pasture. We bedded the small children down in our bed and the bunks. Those men who had horses left to help hunt them, and the people who were in cars left. The women and I had conversation.

By daylight all had been found except one team belonging to a couple with ten children. The father and our most adept tracker circled the pasture and found where the team had jumped the fence and headed for home. They found them about half way home. They all ate breakfast which the wife and I had prepared then left.

Granville and I cleaned up. We got to bed about ten o'clock Sunday and declared we were going to sleep the rest of the day. This followed the rule of all such definite declarations. At noon there was a knock on our door. Granville hastily dressed and went to the door. There stood Stanley Coffin, his wife and daughter Margaret. They had driven up from Yakima to tell Granville that beef buyers would be up to look at the beef a week from the following Monday.

After dinner Granville and Stanley went out to look over the beef that were half in the beef pasture near by and we women visited. They stayed the night and left early the next morning.

The buyers were there by daylight Monday. They looked over the animals and offered ten dollars a hundred delivered and boarded in Ellensburg. One man was left to top them. At four a.m. Wednesday, the men all left for Ellensburg to drive the beef the twenty-six miles. At noon I met them at the foot of the mountains with a hot lunch, then I went on to Ellensburg for supplies. I stayed all night and returned Thursday morning.

About three o'clock that afternoon, I heard a knock at the door. There stood a very tired looking man. His cloths were dirty and he looked as though he hadn't eaten for a long time. He said, "Is your husband here?"

"No." I answered. "He is just out in the pasture."

"I haven't had anything to eat since last evening. I got lost up in your mountains."

"Come in," I said. "I will fix you some food."

Just then, the foreman of our company's sheep outfit rode in. The most welcome sight all day. I had a stew on the stove ready to warm up. I made some biscuits and opened a can of tomatoes and heated with butter, sugar and stale bread. This with coffee made a very satisfying meal.

The three of us sat around and talked. Just when I had given them up, I heard the tinkle of a spur rowel out toward the corral. Granville dismounted and came into the house. Another supper was prepared. After supper, we sat around and talked about the trip. The guest told our fortunes and talked very happily.

The men went out to the bunk house to bed except the foreman of the sheep outfit. He called Granville out on the back porch and told him that he had heard that there was an insane man who had escaped from the State Hospital in Medical Lake. He was thought to be in those mountains, so knowing I was alone there he had ridden over. I shall be eternally grateful. They decided to let things go as they were since everything was quiet.

After an early breakfast, the man asked the shortest way to the Columbia River. He thanked me for the wonderful way I had treated him and said he would never forget me. Granville directed him and one of the men watched to see if he started on the way he was sent.

In about an hour the sheriff arrived from Ellensburg with three deputies. Granville described the man and they said he was the one they were looking for. He had killed his wife and was sent to the asylum. In making his escape, he had killed a man who tried to stop him.

The Sheriff turned to me and said, "Madame, you were sure using your head when you fed him and gave him a bed to sleep in. No telling what he might have done if he was opposed."

He was caught between the Horn Ranch and Columbia River Siding by the two sheriffs from Ellensburg and Wenatchee. Next morning the cutbacks from the beef were turned back in the beef pasture until fall when they would have another sale. Granville picked out a fat heifer calf that had missed weaning and they butchered her. In the absence of refrigeration, they hung her high in one of the tall pines where it was shaded. No flies went that high.

The next morning after it was cooled, they wrapped the quarters in a tarp and laid it close to the spring. This process was repeated about ten days. Then it was cut up. It made some of the best meat I had ever eaten.

Geraldine and her twin brothers arrived at Mountain Home with her maid. A Swedish girl, who had always lived in the city, accompanied them.

We moved to the smaller cabin for a kitchen and our bedroom became a large army tent. This gave the others the large cabin. Secretly I liked the new arrangement much better. It put us in the timber and closer to the corrals where all the action was. The stove was larger and better. The oven cooked fast on the top and slow on the bottom. I could bake pies and biscuits on the top of the stove then the rest in the oven. I never quite figured how to brown them on the top in the other oven.

Another spring led us to a trough where the horses in the holding pasture came up for drinks. We rigged up a place there for doing my laundry. A boiler was placed on a fireplace and running water was readily available

We needed fruit for pies. Granville said I might go down to the Horn Ranch. He sent one of the new hands, a young lad about seventeen down with me. We took a pack horse and boxes that would fit in the saddle bags. As usual I was riding Casey. Before I left, Granville reminded me of the lay of the land and my horses homing instinct.

Amy Wyatt, a friend, was staying in the cabin with her daughter and husband. She helped us pick berries and prepared us a fine meal at noon. In the afternoon we picked some early green apples and some Thompson Seedless grapes. The temperature was about 120 degrees so while Amy and I rested in the shade, Keith went down to the river to cool off.

About three o'clock, we started back up the mountain. With a loaded pack horse and in the heat our progress was much slower going back. It was nearly seven when we got to the Sheridan Place about half way home.

We stopped long enough to get a drink of water then went on. The sun had set and dusk was descending when we left this ranch. It was the first time I had seen it and I would have liked to explore, but decided to go on. Soon it became dark and there was no moon. Keith became alarmed that we were lost. I said, "Granville said to give the horses their heads and they would take us to Mountain Home.

He followed a short distance but worried aloud. After we had traveled about an hour he said, "I'm sure we are going the wrong way."

"We are still climbing. Granville says to keep climbing and they will take us home."

Later Keith said, "I know we have passed Mountain Home. I'm going to turn around."

"I'm not," I replied. He turned and went a few feet. I went on. He could not stand being alone so followed me.

Shortly I heard a voice. I called and our horses speeded up. We arrived at an open gate right across the road from the pole gate leading to Mountain Home.

Three of the cowboys were there to meet us. They said they tried to get Granville to come with them. He went to bed instead. Said he knew I would give Casey his head and Casey would bring us home. "Besides," he said, "It never does any good to search in the dark. I will wait for the moon to come up." The next two days I made pies and canned raspberries and green apples. We gorged ourselves on grapes.

It was time to gather the horses. Early Monday morning, Granville and I set out to locate the herd. We rode down a canyon I had never been on. We saw a number of small herds of horses, most of them carrying the two-bar brand on the left hip. There were some carrying other brands.

In the meantime the men were notifying surrounding ranchers we were going to gather horses on Wednesday. This canyon was much less rugged than the ones I had been in before with less timber and more parks.

We did not like to carry sandwiches when riding through the timber as everything we carried had to go into saddlebags. At noon we stopped by a flowing spring. Granville produced two cans of sardines and a small package of soda crackers. That with the icy spring water made a satisfying luncheon. We rested about one half hour.

We mounted, climbed a hill to the east, and two hours later looked down onto a beautiful valley with an orchard at the mouth of the stream. Several small bands of horses could be seen up and down this valley. Looking upstream Granville showed me a clump of trees that surrounded Mountain Home. Elevation there was slightly over a mile. We descended to the ranch. There were large hay meadows, grain fields and a family orchard with plums, apricots, peaches, apples and grapes. Yellow transparent apples were just ripening and made an excellent dessert for the meal we had eaten. We stopped at this house to tell them of the coming round up. They invited us to stay the night but we left shortly for the Horn Ranch since we had only Tuesday to comb the rest of the lower range.

We headed toward the base of the same hill but how different it looked from this lower altitude. It looked like solid rock and curved with the Columbia River, ending in a point near this ranch.

In a short time we were climbing a very steep hill and I did not know how to adjust my weight. I soon learned why Granville had shortened my stirrups before we started out and I threw my weight on the stirrups leaning forward on the horse. It was much easier going, on the horse as well as me.

We came finally to a more level place but the trail was only about eighteen inches wide. The Columbia flowed about 500 feet below us. Looking down from this point, the section of the river was shaped like a cow's horn. This was the reason for the name of the ranch.

"Don't look down." Granville commanded. I had seen all I wanted so I followed instructions. There was about a mile of this type of trail, then a slight widening. We stopped our horses to let them blow and dismounted. It was fairly level here and we enjoyed walking and leading our horses. We had walked about fifty yards when Granville climbed on his horse but told me to still walk and lead mine. I was quite indignant, but I did as I was told.

Around the bed the trail narrowed and slanted down to a shale hill that raised about twenty five feet above us and sloped steeply down to the river. Our trail sloped gradually around the hill above the river. It was very narrow. Sometimes watching them I thought both horse and Granville would slide down. I was sure now that I would have been safer on the sure footed Casey than on my own hesitant feet. There was about one half mile of this when we came out on solid ground and below us lay the beautiful, wonderful Horn Ranch. Both horses started to trot and we were soon going into the coral there.

The sun was just setting. One of the most spectacular sunsets I have ever seen was over the Columbia River from that height. The blue sky carried all the colors from blue to pale lavender, rose, bright red and gold reflecting back in the broad Columbia.

We were more than ready for the wonderful supper Amy Wyatt had prepared for us. We did not linger long after the dishes were done, but went to the main house to our beds. Leaving early we arrived back at Mountain Home about two p.m. next day after Granville was satisfied he had spotted most of the leaders and their particular harems.

One thing I could depend on when we went on one of these rides was that they would have twenty-four hour beans when we returned. The cook dug a hole about two feet deep. He made it smaller at the bottom than the top. A fire was laid under a grill that was held up by the slimmer wall. The beans were set on top to come to a boil; the beans are fully seasoned with ham, salt, pepper, brown sugar or molasses, ketchup and dry mustard. As soon as they boiled and a good bed of coals were formed, they covered the Dutch oven with the heavy cover. Then a sheet of tin that extended beyond the cover or other fireproof material was placed over that and the hole was filled in with dirt. The next evening the most delicious beans could be served.

A tough piece of beef could be treated the same way and would be as tender as a baby's smile. A cabbage slaw plus an applesauce made us a hearty meal. I found that the man who did the cooking in my absence had two Dutch ovens of ham and beans started for dinner Wednesday.

About five o'clock, three ranchers arrived. By early morning there were twenty men plus our own outfit. The men who were not gathering and saddling horses fell to and helped prepare breakfast. In the four years I was there I never chipped a stick of wood nor built a fire unless I was alone. After eating four of the men did up the dishes.

As day broke, the men left for the roundup. The din ceased and that was the quietest place on the face of the earth. I peeled potatoes ready to boil. Then I baked eight pies. I could put four in this oven at once. I made a large salad of canned peas, carrots and cheese and sat it in a cooler that had been made from a five gallon jar covered and set in a shady spot next to the spring. We kept our milk and butter in a similar jar. I had a large pot roast of beef on the stove in a Dutch oven to make meat in addition to the ham.

Our wrangler boy Bud had made a trip to Columbia River Siding, so we had plenty of bakery bread and other supplies. He had taken two pack horses.

The young people in the other cabin had invited me over to watch the horses arrive. At near one o'clock we heard a sound like thunder advancing. We all rushed out to see Granville on Casey Jones leading the remuda. Anyone who has not seen a herd of wild horses run through a lane to a corral cannot imagine the beauty of this scene. Heads up, muscles rippling and every inch of the animals alive, they followed the lead into the coral.

I hurried back to get dinner on the tables while the men washed up. Such excitement. Every man had picked out one or two favorite ponies. There were about thirty seated at the table. As they ate, stories were told of their experiences in gathering the horses. One man, Len Rickman, who had formerly worked for Granville was working with another outfit at this time. He was the butt of many jokes.

A new man was in our outfit. He came from Canada. That was the name he was called by, said he was a widower with one daughter, Mary, who was cared for by her grandmother in Canada. He was a good rider, understood stock and seemed very anxious to learn the general characteristics of the country.

What a personable young cowboy. I watched the young man walking toward the cabin with Granville. "Alta, this is Hugh Bell. He is going to be working here. He had been hired to break horses as well as general range work. He was in his middle twenties, medium size and complexion. He was long legged, slim hipped and very graceful. His voice was pleasant and low pitched, more refined than the average rider. As the summer moved on, he made a top hand. He understood handling the livestock, taking excellent care of his string of horses. Hugh handled his string of broncos very easily, taking care to keep them from ever bucking. By late summer his string was gentle.

He left the outfit as the time of the Pendleton roundup neared. We did not see him again until next spring. Everyone was very glad to see him since we were very short handed. Hugh said, "I received word last fall that my wife was very ill. I went back to Canada. Helen died shortly after my return. I couldn't stay there any longer so I left Mary, my three year old daughter, with my mother. Here I am!" Hugh seemed to be happy with his work. He was still his charming self, seeing that everything he did was just right.

My baby was due in the summer and he was especially helpful, seeing that I did none of the heavy work. However, when everything was quiet, he became very despondent. Often he went out for long rides by himself. When the others were gathered round or playing cards, he became very moody.

The snow had gone on the foot of the mountains. The grass was greening around the Sheridan Place. It was time for us to follow the pasture and move up. I was glad because it was getting hot at the Horn. I loved the higher altitude and the evergreen trees. Now the work began in earnest.

The range horses needed attention. Riding the range became an everyday occurrence. Toward the last of April, Granville got worried and insisted I should go to Spokane until the arrival of the baby. Bea was born June thirteenth. Three weeks later, I bade goodbye to my folks and returned to Ellensburg, where I was met by Granville. What a great homecoming. They had now moved to the summer cabin at the Mountain Home. What a beautiful place it was. How I enjoyed getting back. I noticed tenseness in Granville. I asked him why he never rode my horse Casey Jones. He said, "He got winded."

"How?"

"Hugh took him out when I was away and winded him."

Casey was the fastest horse on the place. He was always used to leading the range horses into the coral.

"Hugh is still working him. Why didn't you let him go?"

"I needed him here right then."

"How unlike you." I thought, but didn't say.

Granville had to make a trip to the Quincy flats to pick up some stock that had been missed. While he was gone, Hugh came to me and asked, "Did Granville leave his six shooter here?"

"No!"

"Then does he have a rifle here?"

"Yes. Why?"

"One of the range mares stepped in a hole and broke her leg. I need to take care of her." I let him have it. Later, Hugh brought the rifle back and hung it up.

"I cleaned it," he said.

That afternoon, when no one except Joe Helton, an old man who had worked for Granville's father was there, a neighbor rode in.

"Stanley Coffin called and said an emergency had arisen. He wants Granville to call him." I rode the five miles over and called Stanley. I told him of Granville's whereabouts.

Stanley said, "I will contact him there. Could you tell me if a steer calf was branded on the right side instead of the left?"

"I am sure there wasn't. Granville is always very particular about his branding."

Granville came home about midnight. He left early the next morning for Wenatchee. Later that morning the sheriff picked Hugh up. He and another man had been picking up steers. This man had been receiving them and shipping them to Canada. Granville told me he had been suspicious this was happening since he winded Casey. He kept Hugh on so he could watch him more closely. He was convinced of it when he destroyed the rifle.

Inside of two weeks they were tried, convicted and sentenced. They were to serve from ten to twenty years. Hugh threatened Granville's life.

Shortly after we arrived at the feed grounds the next winter, we heard that Hugh had escaped. A few nights later we saw lights playing on the camp from the hill above. There was only one way down to the camp and it led through Kelley's yard. We were sure Hugh was looking for a way down there. After breathlessly watching this for about a half hour, the lights started to dim. Suddenly bright lights came on and took off cross country, then disappeared. We did not sleep the rest of that night. Two days later, Stanley got word to us that Hugh had been captured in the vicinity of Beverly. Beverly was twelve miles from our camp.

A few weeks after that, a Wenatchee paper was sent to us giving an account of his death following surgery for acute appendicitis. I did not think I was capable of being glad to hear of anyone's death. I surely breathed a sigh of relief at his.

We had moved from Washington in October. This was to be our first

Christmas in our own home. Everyone was excited. The children went to bed early and awakened early in the morning. They were delighted with the gifts in their stockings and on the tree. By dinner time they had gone back to playing with their old, familiar toys as children usually do.

Just before we sat down at the table a buckboard stopped by the door, an Indian man and two women got down. They wished to get warm. We asked them to eat with us and the man, who had said his name was John, said, "The ladies are hungry and would like to eat but I do not care to eat." Of course a plate was set for him. As is usual when a man says he is not hungry, he ate twice as much as anyone else. He was the only one who accepted a second helping of mince pie.

During the meal John told us about a building we'd seen earlier in the day. He said it was a steam bath house. They heated rocks, then poured water over them and steamed themselves, then ran naked into the icy water. This toughened them so they did not catch cold or feel the cold.

Newborn babes were dipped naked into the cold water, then wrapped in blankets. This was supposed to toughen them. The sheriff tried to stop this practice, but since they did not use the same place twice, it was almost impossible to catch them.

John also told us that they had belonged to a tribe of Kootnai Indians in Canada, but one of the chiefs had been dishonored and their tribe had fled across the line to save their lives. They were never fully accepted by the Indians in this country so they had no tribal or reservation rights. They could not go back to Canada.

They had built houses on some government land and had been allowed to stay. They worked on ranches and in logging camps but they were still just mavericks.

When they left they gave each of the children a hand carved wooden doll which they enjoyed greatly. Also one of my mince pies was missing from an outside cupboard.

We profited greatly in understanding from the tribe's friendship. Although we served many dinners we received many services from them. They brought us fish, huckleberries from the mountains, high bush cranberries, and much folklore.[77]

Myrtle & Arthur Carroll

WINTER 1936

Wyoming winter wind can find you anywhere, even holding close the edges of your blankets as you sit in front of the fireplace with doors and windows secured and draperies shut. It stabs through your clothing; it bites at your hands; it sucks the heat from your body as you speak. Warm fires, warm friends, warm hands pull us to safety and free the ice from our hearts even as it clings to our boots.

The table, extended to its full eight foot length, was laid with a white cloth barely visible under platters heaped and steaming. A sideboard was covered as well, overflowing with breads and sweet rolls and butter and jams and jellies and pie hidden under embroidered dishcloths. Christmas was a big affair for the Carrolls even in hard times.

Arthur and Myrtle, Bob, Art, and Harold gathered with Granville and Alta and eight of their twelve children. Also present were Jewel, Orpha, Ednis and their families; all of George and Ella's children, and most of their grandchildren were together crowded under one roof.

The cousins played cards. Harold was sixteen and sure of himself, winning easily. "I'll deal," he volunteered, reaching for the cards.

"Not this time, little brother." Art had the cards first and was already shuffling. Art was nineteen and scheming, determined to find a way to continue in college. The University of Minnesota offered a good science program and was close enough to home to make Myrtle happy, but far enough away to make Art happy. U of M was also in his price range after much figuring and finagling of costs and requirements with Granddad George. Art planned to lease the local Shell Oil station in order to earn enough money for college.

Bob folded. He held a triceratops bone, turning the fossil over in his hands slowly, imagining its age. He had returned home for Christmas an officer. He'd been honorably discharged to accept a commission as Second Lieutenant of infantry in December of 1936. He had achieved a dream.

Hitler had also achieved a dream. He'd sent German battalions into the Rhineland and marched his troops brazenly into French territory. Emboldened by a quick victory, he devised to add Austria to his empire. Perhaps he rubbed his hands together in true conquer-the-world glee as he considered his next steps.

THE LINE RIDER
By Bruce Brockett

When you've been in a line camp alone for months
And ain't seen much but snow,
When the flour gets short, and the beef's all gone
And the coffee's gettin' low,

When you ain't seen a man to talk to
Since you left the ranch last fall,
When a horse is all you've argued with
And he can't talk at all,

When the gray wolves howl 'round your camp at night
A mournful, lonesome tune,
And the crazy coyotes yip and yap
At the cold old winter moon,

When the cold frost snaps the cottonwoods
And you hear them pop all night,
When the fire goes out and your shack gets cold
And you pull your soogan tight,

There's a sound that will thrill you through and through
When you're layin' there in the dark,
The whistle of a man—you can hear it plain
Above the coyote's bark;

You prick up your ears 'cause your ears are good
Even if your voice is stale,
And you try to figure out the gent
Uh whistlin' up the trail.

Slim whistles "Annie Laurie,"
Nig likes "The Bird on Nellie's Hat,"
While old Dutch sticks to "Redwing"
Wherever he is at.

Then as the sound gets closer
It's good old Dutch, you know
'Cause "Redwing's" the tune you're hearin'
Floatin' across the snow.

You're out of bed in a second
And light up a fire right away
'Cause you know the old boy will be hungry
If he's come from the ranch in a day.

You put on the pot for coffee
And put some sourdough in a pan
He can eat some corn, too, you reckon,
So you get out your very last can.

Then you sit down by the fire and wait
So you can act like you ain't glad at all
To see the first man you've looked at
Since you left the ranch last fall.[78]

Myrtle, Bob, Ella, & George G., & Art wearing the hat. Notice the strange hat Ella is holding.

PART II

Peter Carroll Family about 1904.
Back row: Robert E, Myrtle L, Arthur C;
Front row: Peter, James, Martha Atherton Carroll,
Margaret Ruth

CHAPTER 15

EMIGRANTS & UPRISINGS

"My master is a great tyrant," said a Negro, according to a popular quip of the day. "He treats me as badly as if I was a common Irishman." ~ Noel Ignatiev

JAMES MCCARROLL STEPPED ONTO ELLIS ISLAND IN 1861. HE'D LEFT WIFE AND CHILDREN BEHIND IN GLASGOW, SCOTLAND and a past unknown to his descendants, or the emigration officials, except his place of birth—Ireland.

Ellis Island sits in the middle of New York Harbor just off the New Jersey coast; the Statue of Liberty casts her shadow from a tiny nearby island. James joined seekers of all sorts being herded in droves into a giant, Victorian style central building that faces the vast Atlantic.

When you look up from the main floor, the ornate ceiling can barely be seen three stories above. The architecture was designed to intimidate, letting the minions know the moment they entered its portals, "you are a speck—the Institution is Omnipresent." Iron railings marked the passage and the hopeful shuffled through the walk-ways in single file. Hundreds of people stood in processing lines. hoping their English was good enough they would understand the questions thrown at them, hoping they'd pass the medical exam, hoping the tricks they'd devised or the bribe money they had would be enough, hoping to see family soon.

James McCarroll's history is even more difficult to trace than George Carroll, Sr.'s. Their sons, Grandpa Peter born in 1860 in Scotland, and Granddad George born in 1860 in North Carolina, seem to step out of a veiled past.

We know that James McCarroll was born in Ireland, but we don't know exactly where nor his parents' names or religion. We know he married, attended church and christened his children in Scotland. And we know he immigrated to America and later sent for his children. Other than that, James McCarroll remains a mystery.

How did he travel from Ireland to Scotland, and why? Since earliest times, the Gaels of the British Isles intermingled. People swam, sometimes successfully, across the North Channel, or Sruth na Maoile as the dangerous 22 mile swath of water is called in Gaelic. It separates northeastern Ireland from southwestern Scotland—or in fact it joins them. The people on both sides of the channel are of Gaelic roots, as are a majority of people of Ireland and Highland Scotland.[79]

James McCarroll may have come from County Leitrim, Ireland according to Merle Harris' genealogical research. The chief industry there was a tobacco firm which manufactured Sweet Afton cigarettes and it was begun by Patrick James Carroll in about 1824. A tobacco grower in Ireland would have had some difficulty making a living as tobacco was considered an illegal export under English law at that time.

It is likely the McCarrolls were Catholic and speakers of Gaelic. It is possible that the McCarrolls were involved in the newspaper trade or politics as skills and occupations were typically passed on from generation to generation. Some of James McCarroll's children and grandchildren became orators, newspapermen, and politicians.

As Catholics, if journalists, their lives might have been in jeopardy. Any newspaper they contributed to would have to have been an underground paper. Catholics in Ireland at this time had no rights to own property or goods over a certain value, no right to education, and no political voice. "Catholics were not permitted to vote or serve in Parliament or hold public office, or live within the limits of incorporated towns; they were forbidden to practice law or hold a post in the military or civil service. Catholics were forbidden to teach, attend university, or educate their sons abroad.

"They could not take part in the manufacture or sale of arms, newspapers, or books, or possess or carry arms. No Catholic might own a horse worth more than five pounds. They might not buy, inherit, or receive gifts of land from Protestants, nor rent land worth more than thirty shillings a year, nor lease land for longer than 31 years, nor make a profit from the land of more than one-third of the rent paid; No catholic estate could be entailed but instead had to be divided at death among all the children." Inheritances were prevented and wealth could not accumulate in the hands of the Catholic in Ireland during this time. [80]

Born just twenty years after the failed United Irish Rebellion of 1798, James would have grown up hearing stories of the Society of United

Irishmen headed by Theobold Wolfe Tone. Wolfe Tone launched an uprising in order to right the wrongs of English oppression against the Irish. The uprising failed when the Irish were defeated by General Cornwallis. "The defeat is ordinarily ascribed to unfavorable winds that kept the promised French aid from arriving on time, but a more important reason is the failure of the united Irish leaders to link their demand for a democratic republic with the struggle of the Catholic peasant for land."[81] Over one hundred thousand took part in the Rebellion, and almost twenty-five thousand died. Then Great Britain eliminated the Irish parlament.

The generation without an official voice raised even angrier, louder children than their parents, and they began their own uprising. The Young Ireland movement started when Daniel O'Connell called for the repeal of the 1801 Irish Act of Union with Great Britain, an Act that effectively subordinated the irish ecomomy. The fighting Young Irelanders were mostly middle-class writers and intellectuals rather than warriors. They wielded weapons occasionally but generally wielded mighty pens, but not mighty enough. Their oratory and armed action did not reduce English oppression.

James McCarroll was eleven in 1829 when O'Connell campaigned unsuccessfully for a Catholic voice in parliament. It seems logical, from an American perspective today, for the Catholic Irish to want a representative in the Government that made decisions about their lives and livelihoods. Those with eyes saw that decisions made by British Parliamentary members had led to poverty in Ireland; poverty had led to starvation; and starvation led to desperation. The French sociologist, Gustave de Beaumont, visited Ireland in 1835 and wrote: "I have seen the Indian in his forests, and the Negro in his chains, and thought, as I contemplated their pitiable condition, that I saw the very extreme of human wretchedness; but I did not then know the condition of unfortunate Ireland...In all countries, more or less, paupers may be discovered; but an entire nation of paupers is what was never seen until it was shown in Ireland." The next year, in 1836, people in England celebrated Christmas for the first time as an official national holiday.

"The Irish phrase 'An Gorta Mor', meaning 'The Great Hunger', is regarded by some as being an accurate description of years of hunger, which were not simply caused by food shortages...[because] between 1845 and 1852, large volumes of food were exported from Ireland..." During this same time, "at least one million people died out of a base population of over eight million people." Because of this Great Hunger, "Ireland is the only country to have a smaller population [in 1997] than it had in 1840."[82]

In addition to mass starvation, "a half million were evicted from their homes during the potato blight, and a million and a half emigrated to America, Britain and Australia, often on board rotting, overcrowded 'coffin ships.'... Dr. Kinealy's research proves beyond a reasonable doubt that there was sufficient food in Ireland to prevent mass starvation and that the food was brought through the worst famine stricken areas on its way to England. British regiments guarded the ports and warehouses in Ireland to guarantee absentee landlords and commodity speculators their 'free market' profits."[83]

Poverty in Ireland and its impact increased unabated due to the laissez faire economic policies of the British government. Relief official Charles Trevelyan took a particular glee in allowing the Irish to starve to death for the good of The Market he held in reverence.[84]

"In 1848, Trevelyan published his own account of Famine relief. This was the only written account of the Famine produced by a senior relief official. He employed a moral and providential framework in which to place the Famine, which he described as 'the judgment of God on an indolent and unself-reliant people, a people, moreover, who liked to 'make a poor mouth'. He further asserted that no government had ever done more to alleviate the suffering of its people."[85]

Trevelyan and leading political economists of the day "doggedly argued for less government intervention and more free trade." They asserted that "the Famine crisis made it all the more important that the principles of political economy should be applied to Ireland. Any relaxation, however nobly motivated was a 'killing kindness.'" Even though Britain demonstrated "willingness to intervene in the marketplace when it chose to do so, unhampered by ideological or moral constrictions...For example, Britain went to war with China (1839-42) to force it to accept Britain's trading policies, including the sale of opium to the Chinese population."[86]

Historian Kinealy notes, "Ireland had one of the tallest, healthiest and most fertile populations in Europe" prior to the Famine. "It is only now, as Ireland emerges with a distinctive and positive identity within Europe, that Irish people throughout the world have been able to come to terms with the impact of these [Famine] years and define what it means for their culture and history."[87]

We know little of James McCarroll, in part, because oppression impacted the flow of information as well as food. Families were torn apart, religions abandoned, lives lost. And sometimes from such trouble comes good; love from the ashes of war, joy from the pain of birth.[88]

We know from genealogical records compiled by Merle Harris that James McCarroll married Catherine Sharp when he was only nineteen years old, and that James was dubbed an "Irish defectionist."[89] This was someone who left the Irish side for the British side, or a Catholic who married a Protestant. "By converting to Protestantism a Catholic son could dispossess his father and disinherit all his brothers."[90] And maybe save his own life in the process.

It was told that the first four children born to James and Catherine died young, and that the next four children born to them were given the same names. It was typical of Irish families of the time to name their first-born son after the paternal grandfather and their first-born daughter after the paternal grandmother. The second born boy and girl were named after their maternal grandparents.

James and Catherine's oldest daughter, Cathleen, married in 1858 in Glasgow. Their oldest boy, Walter, was only sixteen years old when James went to America in 1861. John was twelve, James Jr. eleven, Margaret nine, Jeanette seven, Robert three, Peter (Bob's, Art's and Harold's Grandfather) not yet two, and Thomas was still growing in his mothers' womb when their Pa left Scotland, never to return.

He was among the almost one million Irish—about twice the total for the previous two centuries—who sailed for North America in the 1800's. They didn't conform to the stereotype; many were Protestant, spoke no English, and came with plenty of money in their pockets. Some money had to be available in order to book passage to America. Many, like James, had given up their Catholic faith and had become Protestant somewhere along the way.

Arriving in America brought some surprises to those who thought they'd left prejudice far away on distant shores. "American prejudice was of a slightly different variety. The label 'Scots Irish' was unknown in Ireland. Many protestant Ulsterites had intermarried with native Irish women and their children grew up as simply 'Irish.' The shared distinction between Irish and Scots-Irish developed in the United States in the last half of the nineteenth century. After the great influx of Irish immigrants … the Scots-Irish insisted upon differentiating between the descendants of earlier immigrants from Ireland and more recent arrivals."[91] Bigotry is often greater between longtime residents and more recent immigrants from the same cultural heritage, as the established folk try to escape the stigma of prejudice being leveled upon the newcomers.

In an effort to avoid this prejudice, James, like many other "Micks," dropped the Mc from his name when he left Scotland in 1861, the same year George, Sr. was wounded. The American Civil War was in full force.

James Carroll made his way from Manhattan Island into rugged Pennsylvania mining country and hired on as a miner. He saved and sent money home to buy his family's passage as soon as the war was over.

But Ireland was in the throes of famine and disease, and life did not wait for the war to end or James' financial readiness. His oldest daughter,

Cathleen, died in 1862 in Glasgow of a contagious fever, probably smallpox. Her husband and children, grieving her loss, presumably stayed in Scotland. And James' wife, Catherine, died in 1865, probably of the second smallpox epidemic that hit Scotland. She is buried in Glasgow. James' and Catherine's oldest boy, Walter, married in 1865 and escorted his bride and younger siblings to America, including Peter Carroll, later to be known as Grandpa Pete.

SUMMER 1937

Peter Carroll's cabin above Story, Wyoming was filled with the sounds of the whispering pines that encircled it, and the fragrant smell of the icy cold stream that flowed down from the snow melt right behind the simple log structure. You can smell its sweet, mossy scent from the porch as it tumbles over the smooth stones. It was also filled with memories—memories of the many gatherings, friends and family, that had filled its cozy rooms with laughter and music over the years. Martha (Matty) used to smoke a little tobacco pipe on the good sitting rock near the stream.

Pete had been alone since 1919 when Matty died suddenly of a heart attack. He'd left the lonely stone cabin they'd built together at his children's bidding and the urging of a persistent case of bursitis, and built a new, modern log cabin nearer to the stream.

If you mentioned Peter Carroll's cabin, someone would likely say, "Do you remember the time when…" Harold was there on this fine summer day chopping firewood, and doing chores, and just piddling around. He rubbed his head in memory. The smack to the back of the scull growing up was always unexpected.

"What's that for?" was the usual anguished question, with the ritual reply of, "that's for nothing. Start something."

They'd all been victims of Grandpa Pete's pranks, but usually good outcomes favored children. Peter was a connoisseur of humor, even if others weren't always appreciative of his particular brand.

Harold fondly remembered a sly and mischievous Grandpa Pete and the small alliances that grew between grandparent and grandchild. For example, Pete had been warned not to give the children sweets by their mother, so he stopped by the candy store in Sheridan and contemplated for a good long time before selecting lemon drops.

"Good Morning, Mr. Carroll." Charles was always formal to his customers, even those he'd known for years.

"Good morning yourself, Mr. Griffin," answered Pete. He leaned heavily on his cane and scowled as if making a tough decision. "I'll take a half pound of those lemon drops," he said matter-of-factly.

"Again?" Charles asked, incredulous. "You've got no imagination at all." He scoffed and shook out a small white bag and ladled in a portion of yellow candy just a bit larger than a half pound.

Later, Art and Harold, just boys, were playing with their cousin, LaVina, at the cabin as usual. All three gathered around Grandpa Pete. Kids and animals always gravitated towards Peter Carroll.

He sat on the stump outside the hand-built rock structure constructed when he and Matty'd first moved here all those many years ago, and sucked on his pipe. Little puffs of smoke wafted heavenward. Peter pretended to gaze thoughtfully at the sunset. He doesn't smile, but the children can tell from the glint in his eye that he's playing with them.

Peter now skillfully extracts from his magical pocket just one large, sugary sphere each for Art, Harold, and LaVina, slipping it to them craftily so no one else will notice.

There was little verbal conversation between Pete and the children, little was needed. Not that Pete was a quiet man, far from it. Peter's brogue became thick when he was angry at anything. Irish idioms escaped his mouth like he'd grown up there when he'd never set foot in Ireland. The language of his father, James, came out of his mouth naturally.

At Grandpa Pete's cabin, children were taught to be proud of their Irish heritage, but Peter could be whatever suited him. He might call himself Scottish when talking to one person—he was born in Glasgow after all—and Irish when talking with another, as his father claimed roots in County Leitrim, or he could be Scots-Irish if he felt like it.

Peter had become a senator in part to make proud his hardworking father, James. Pete fought for the rights of the little guy— his fellow miners, his neighbors, his family. Eight hours is plenty for a man at hard labor; he knew this was true. He'd fought to end fourteen hour days. And he won his battles, many of them, and sometimes they grew more heroic with the telling. A skilled fly fisherman, Pete could tell a fine fish story.

When he told tales to his grandchildren, they included stories about his father, James McCarroll, and his murky Irish past. He liked to point out to Bob how much he looked like his great-grandfather. Pete also told dark and scary mining stories, claustrophobic tales of the subterranean. He mostly avoided ruminations of his childhood memories, the untold tales of growing up in the post-civil war mining communities of Philadelphia.

Pete himself was curious about his father's history, and sometimes aloud. "In 1836, on a skirmish near Belfast City, James was wounded. The whistling sound was so loud and time was moving so slowly it almost seemed as though he should have been able to get out of the way. The impact of the bullet on his right shoulder knocked him back several feet and he fell into the underbrush, saving him from the next round headed in his direction."

"What happened next, Grandpa?"

"Well..." Peter started slowly, warming up to his topic:

His friend, Michael, dragged him deeper into the bushes and covered them both with tree branches as the British searched for them in vain. And James, though bleeding profusely from his wound, struggled with Michael's help to the little cabin a few miles deeper into the woods.

Catherine had come from Glasgow to Belfast City to visit her cousins. Her parents were of Irish heritage, but she was born and raised in Scotland. She was unprepared for the distress she encountered. The men had disappeared days before her arrival. Frightened and angry, the women of the household waited at home wringing their hands, considering what to do.

Suddenly, a man burst through the cottage door. He was covered head to toe with blood, his clothing torn and ragged, yet he entered the drawing room proudly."

"Michael, are you all right?" his sister Margaret rushed to him, looking for the cause of the blood. Catherine had not seen Michael since they were toddlers and rough introductions were made. "This is your cousin, Michael," Margaret said, nodding curtly to the young man shoving her assistance away.

"I'm fine, I'm fine." He glanced up at Catherine, "Pleased to meet you." Michael turned abruptly to the fawning Margaret. "It's James. He's been shot. I left him at Murphy's cabin. He needs help."

The girls quickly gathered medical supplies and followed Michael on horseback out of the city and through a wet and rolling countryside to the shabby cabin deep in the forest. When James McCarroll met Catherine Sharp, they fell instantly in love then and there.

Other times Peter would tell of his Grandfather, Walter. Although he'd never met him, Peter knew him from James' detailed complaints of Walter's sour disposition. How he came home to his children after each long day bitter and angry, his misery as relentless as the damp, never-ending rain that season.

"Me father, James, and his brother John were good workers," he began:

The older, thatched roofed home sat on a lovely hill in the Irish countryside. It was warm and smelled of peat and oat bread, and an abundance of poverty and dashed hopes.

The boys, John and James, made sure the animals were fed and the evening chores completed, while parents, Margaret and Walter, talked avidly of the day's events. Margaret walked away from dinner preparation this evening to argue with her husband, and left the chore in the capable hands of youngest child, Caitlin.

"No Catholics are allowed a vote in Parliament!" Walter announced. "Things will remain the same. People have lost their lives for nothing."

"Not for nothing, Walter," Margaret soothed. "Change will come," she pleaded, knowing their talk would now turn to emigration.

"We must make the change ourselves," Walter insisted. The argument continued like the sad lyrics of a poem.

"But return they don't!" Margaret lamented, "the children never return."

"It's our only hope," he insisted.

There was certainly not enough money for passage for all the McCarrolls to leave Ireland. Walter thought it better to risk his sons; they'd have a possibility of earning the money needed.

The original plan was for the boys to go to America where they would find work and send money home, eventually buying passage for the entire family and a new start away from poverty and oppression.

With his dream of Irish independence failed, Walter knew he had nothing to offer his children or grandchildren. No one child could inherit the family business. Thieves were already lining up to snatch the business once the property was broken up. And the shortages were beginning; the potato crop was blighted again this year. There'd be reason to sell at no profit.

The McCarrolls' life was one of poverty with a gloomy future inevitable. No future— a young man might argue. There was nothing left to stay in Ireland for except hardship and famine. It was 1836. James was 17.

James got a job on a ship that sailed along the west Irish coast with the hope of earning enough money for passage to America. Unfortunately, he was taken advantage of on his first night aboard, and what little money he had disappeared like the fleeting Irish sun, money that was scraped together by his family and should have taken him at least to the departure point to America in Westport. All the money gone—squandered or stolen— he was required to work even longer for enough for passage.

The vessel he cleaned eventually harbored in Glasgow and James left his post as swabbie and found a land-laboring job. While in Glasgow he met, fell in love with, and was fully detained by the beautiful Catherine Sharp.

While sailing in Protestant waters, James claimed to be Protestant. It was certainly better than dying young. He gave up his Catholic heritage formally when he committed his future to Catherine, a Presbyterian, and relegated his dream of coming to America to some indefinite future.

Who knows what really happened? In any case, in 1838 in the wilds of Scotland, James the brave married Catherine the fair. And they lived happily ever after. Or at least, for a few years they resided in Glasgow.

Imagine James' great love for beautiful Catherine and his empty pockets. He could not have brought Catherine with him to America. She so delicate and with child, it was such a hard journey under the best of circumstances. How difficult it was to leave when finally he did. In 1861, he said his farewells and braved the harsh Atlantic crossing.

James wrote frequently to his family, sending what money he could. He tried to keep the letters home brief and bright.

---

7 April 1862

Dearest Catherine,

How are you darling? How are the children? I long for word from you. I hope everyone is well.

I have found work and have enclosed what little I am able, more will come in the future. I can hardly wait to see you and the new babe. The city is more than I can describe. There are so many people here in Philadelphia. You will love all the many different ladies' hats.

The American Civil War must end sometime, and when it does, I will surely have the money saved for yourself and the children to make the crossing to America.

How I miss you.

All my love,

James

Peter Carroll

CHAPTER 16

MINERS, RAILROADERS, & ANCIENT HYBERNIANS

Power always thinks it has a great soul and that it is doing God's service when it is violating all His laws. ~ John Adams

THE FIGHT FOR IRELAND AND THE FIGHT AGAINST OPPRESSION DIDN'T END AFTER EMIGRATION. IMMIGRANTS and their sons and daughters sent money and arms back home to Ireland. The Fenian Movement called the next angry generation into action, seeking an Irish revolution "sooner or never," and carried on the ideas of the Young Ireland Movement with the help of their family abroad. "Fenianism was strongly supported by Irish emigrants in America. Many gained military experience in the American Civil War," and after the war ended, many returned to Ireland.[92]

Many more Irish stayed, filling the streets of New York, Boston, and Philadelphia with cheap labor. James fought the same battle he and his family had fought across the Atlantic. "In Ireland they were landlords and agents; in Pennsylvania they were mine owners and mine bosses."[93]

James and other immigrants lives were spent working fourteen to eighteen hours a day under dangerous conditions for little pay. The only thing worse than this pitiable existence was the moment a man received his draft notice stating he had to go fight in the American Civil War. Unless he had the money to pay someone to fight in his place or to pay the government—typically $300 to avoid conscription—he could end up, like others, taking orders. James may have joined the throng of protesters in 1863, mostly Irish laborers, who filled the streets to oppose the new draft law.[94]

Powerful people fought hard to keep the struggling immigrant workers in their place. Railroad boss Asa Packer, and coal baron Franklin B. Gowen, worked to pass anti-labor legislation. Then they hired private police to control their angry workers, and their private police were backed by federal troops. "The draft riots were treated as outright insurrection against the U.S. Government that required military force to deal with. The riots were ignited by a policy of exempting the rich from fighting in a war not seen by all as a noble crusade. The riots were led by ethnic Irish and anti-war Democrats, but the Government feared their spread."[95]

Schuylkill County, Pennsylvania, where James lived, worked, and where he brought his children, was occupied by the military for most of the Civil War. Rioting union laborers were the last thing the coal barons and their financial backers in England wanted. Frank Gowen had money and power and used both to take control of the situation. One part of his solution was to run the independent mine owners into bankruptcy, especially those independent miners involved with the Workingman's Benevolent Association (W.B.A.) during the strike of 1871.

Frank Gowen wanted nothing less than the absolute destruction of the W.B.A. and he had the money and power to obliterate them. Step one was to send his Pinkerton hired police to infiltrate and tie the non-violent W.B.A. to the notorious Molly Maguires.

The Molly Maguires were named after a violent group of "protesters" who fought anti-Catholic persecution back in Ireland. Some believed they were members of the Ancient Order of Hibernians. "Centuries of abuses had led to the formation of secret societies across the Irish countryside as the only self-defense agrarian Irish had to combat murder, land confiscation, and religious persecution. These societies accompanied many Irish laborers who found their way to the booming coalfields of Pennsylvania."[96]

The miners had their secret society but it was no match for the unflinching Pinkerton secret police, who also operated clandestinely and outside of the law. The Pinkertons hired James McPharland, an Irish immigrant himself, to spy on the Mollys. "McPharland's orders were to implicate as many unionizers as possible in any crime he could. In many cases, McPharland appeared to 'plant' evidence to incriminate and entrap the miners."[97]

According to Anthony Bimba, "It is clear that the Pinkerton agents did not confine themselves to a spy service nor even to the activities of the agent provocateur. When they could discover nothing and invention would not serve them, they actually committed crime and charged it to the miners." McPharland himself boasted during the trial that he had known of the plan and let the killings proceed to accumulate evidence against the Molly Maguires.

On June 21, 1871, "Black Thursday", at Carbon Country Prison in Jim Thorpe, Pennsylvania, four members of the Molly Maguires were hanged for the murder of two mine bosses. The trial was described by Judge John P. Lavelle as a "surrender of state sovereignty. A private corporation initiated the investigation through a private detective agency. A private police force arrested the alleged offenders, and private attorneys for the coal companies prosecuted them. The state provided only the courtroom and the gallows." [98]

The violence didn't end there. Twenty more Irish immigrants were killed as alleged Molly Maguires and many others lost their lives in the struggle for fair and safe treatment of mine workers and a living wage. "The lead villain was Asa Packer, founder of the Lehigh Valley Railroad, and a deadly enemy of organized labor. The prosecution was a race war as much as a class struggle, and the executions should be contextualized alongside the lynchings that were so prevalent in the American south a few years afterwards."[99] James Carroll had left the fire for the frying pan.

Many whose families come from Schuylkill, Pennsylvania, cannot trace their ancestors because in times past it could be detrimental to admit a relationship to anyone labeled a Molly Maguire. Families separated, feuds ignited, and Irish genealogy suffered greatly. Many descendants are unaware that their Irish roots include a Molly because the information was removed from family bibles and oral histories.[100]

In 1870 James' and Catherine's oldest son, Walter, who'd ensured his younger brothers and sisters had arrived safely in the U.S. just a few years earlier, is said to have met "foul play" and died when he was a young man of twenty-four. Walter had only been in the fabled land of the free for six years before his untimely death. He left a widow and small children behind. According to letters, his widow had conspired with another man to murder him.[101]

Peter was eleven when his big brother Walter was killed. He grew up with the family's talk, or sudden lack of talk regarding the Molly Maguires. This experience may have contributed to Peter's becoming a staunch union supporter as an adult.*

Life goes on and James remarried a woman named Christina Bennet in 1871, giving his children a mother. They headed west soon after, settling

---

*   Peter was a Republican and Union supporter and this was not contradictory at the time.

finally in Trumbull County, Ohio where he became a U.S. citizen on the 5th Day of October 1871.

James and Christina moved to Boone County, Iowa with their children: teenage girls—Margaret nineteen, Jannet seventeen, Jane fifteen; and sons—Robert thirteen, Peter eleven, and Thomas nine.[102]

James Carroll lived ten more relatively peaceful years and enjoyed grandchildren before he died on September 1, 1881 at the age of sixty-two. His gravestone reads: Precious in the sight of the Lord is the death of his saints. His Last Words: Meet me in Heaven.[103]

"These are the twenty Irishmen
Who were executed
As alleged Molly Maguires
If ye find yer kin among them
'Tis then ye may feel the fire...
Bergin, Martin
Boyle, James
Campbell, Alexander
Carroll, James
Donahue, John "Yellow Jack"
Donnelly, Dennis "Bucky"
Doyle, Michael J.
Duffy, Thomas
Fisher, Thomas
Hester, Patrick
Kehoe, John "Black Jack"
Kelly, Edward
McDonnell, James
McGehan, Hugh
McHugh, Peter
McManus, Peter
Munley, Thomas
Roarity, James
Sharp(e), Charles
Tully, Patrick"*

---

* Folk song.

WINTER 1938

It was well past midnight when the sheriff showed up at the door. After hearing his news, Myrtle screamed, "no!" and threw her coat on over her nightgown and robe. Arthur struggled into his boots, complaining about the hour. He kept his mouth shut most of the drive. He was angry at his youngest son for being in an automobile accident at this time of night in the snow, and even angrier that he might be seriously injured.

The group of teenage boys had been drinking, driving too fast, had come to a tight corner and didn't make the turn. Harold didn't realize at first that he was even in an accident, having been thrown clear of the wreckage. In shock, totally disoriented, he did not know both his legs were broken. But when he discovered his buddies were trapped in the car, bloody and unconscious, he walked back to town two miles for help.

The diner was well lit even if the customers had dwindled to a few. Harold approached the nearest man, someone he thought he knew. He tried to explain about the wreck and his injured friends, saying help was needed. "No! Not me!" Harold insisted, stepping back but falling over.

That's when the man shouted, "Close your mouth before you bleed to death!"

Arthur and Myrtle waited anxiously in the hospital corridor for word of Harold's condition, hoping and praying silently, much as James had waited to reunite with his children, hoping and praying for their safe passage to America.

James McCarroll stood on Ellis Island under a dark grey sky, the same gray color of the losing confederate's uniforms, on a fall day in 1865. He saw his son, Walter, standing tall and proud beside his new bride, and James' heart swelled in his chest. He searched his daughter-in-law's pale face. She had become ill on the boat and was still not feeling well and it showed in her countenance. The epidemic that hit Scotland in 1861 had devastated James' family. No one mentioned the fear that she, too, could be suffering from smallpox. But all prayed silently that her cough would go undetected by the authorities.

Standing near Walter, young Peter Carroll also searched the crowd looking for his dad, James McCarroll, whom he barely remembered and had known only from others' talk. He remembered the sound of his voice, his smell, but looked in vain for a familiar face. He stretched trying to see over the many heads, people shoving this way and that. Peter was able to stand his ground in the bustling crowd.

And Peter stood his ground on the streets of Schuylkill. Peter Carroll's experience of life at ten years old in 1870 Pennsylvania was far different from George G. Carroll's life at ten in Bell County, Texas in 1870. Peter left plague-ridden Scotland for post-civil war America and a tough Pennsylvania Irish mining community. By 1870, Peter had grown used to the life of raids, protests, violence, filth, and long work hours for his father and older brothers (Walter, John, James, and Robert), and the dangerous alleyways of the slums he walked each day with Thomas, his younger brother by two years. Some had mothers who provided a buffer for a boy stuck between the rock of the nineteenth century institution and the hard place of the streets of Philadelphia.

Peter often cried late at night into his pillow, missing his mother, Catherine. His older sisters: Margaret, Jannet, and Jane cared for Peter and the younger children as best as they could, and cared for the house, fed the men before they left and after they returned from the mines, and Margaret and Jannet worked themselves in factory jobs with other women sewing garments for the wealthy.

Here, as back home, fighting oppression was a family affair. Walter had heeded his father's advice to avoid the rabble-rousers, just do his job and keep his nose to the grindstone, but he joined with other unionizers when necessary. He was still dirty from the mine, relaxing in the pub with the other abused miners, before going home to wash and bed, and then do it all again for another twelve hours and not enough pay to survive.

His conniving wife was there, unexpectedly, with a cousin he'd never met and

Front Row: Thomas Carroll & Mary Hay Carroll; Back row Gertrude, Lyle, Margaret, James, William, Irene Naomi Carroll Schmoker

they raised a few glasses together. Unknown to Walter, the man was no cousin and was instead her lover and an informant. They drugged Walter's drinks. He stumbled into the night groggily, unsure of himself. That's when he was jumped from behind, and attacked with boards and fists. Thankfully, Walter was knocked unconscious and didn't feel what happened next.

James decided he had the power to act. He couldn't bring his oldest son back to life, but he could escape the mines of Pennsylvania. He could take his family west…

Imagine the moxie required to pick up your family and move across the ocean or across the prairie taking only what you can carry and having just enough money or bargaining power for a few days. Imagine the fear of the unknown completely outweighed by the certainty of a dismal present leading to a more dismal future.

Imagine the kind of moxie it takes to walk two miles seriously injured, your concern for your friends driving you forward. (Or maybe it was alcohol and stupidity that drove Harold.)

Or the kind of moxie it takes to persist with the dream of education, even though the reality was Art had no money for tuition.

Or the kind of moxie it takes to become a paratrooper, throw yourself out of a plane and watch the ground rushing toward you. If you land successfully, you may see the enemy rushing toward you!

Imagine the kind of moxie it takes to gather an army against western Europe, the Soviet Union, and the United States. Or to take Nature into your own hands and mold it like you are God. German physicists had discovered how to unlock her secrets; they'd learned to split the atom. With that, the world changed suddenly— ober nacht.

CHAPTER 17

LITTLE BIGHORN
What has been done in my country I did not want, did not ask for it.
~ Mahpiua Luta (Red Cloud) of the Oglala Sioux

P ETER AND MARTHA CARROLL MOVED WITH THEIR CHILDREN TO STORY, WYOMING, A PLACE OF SPECTACULAR BEAUTY. Not so long ago, it was a crucial trading area for the Crow Tribe.

"The Sheridan area, once known as 'Absaroka', the land of the Crows, can be roughly defined as stretching for a hundred miles in every direction from the City of Sheridan." It was the approximate extent of the Indian trade area and it had a long and turbulent history.

In August of 1811, "A party of Astorians came up through what is now known as Johnson County and crossed the Bighorn Mountains south of the present town of Buffalo. There are some interesting stories to the effect that the Spaniards were in this part of the country in the early part of the seventeenth century—that they built stone houses and arastras and for nearly a quarter of a century sent trains to the south, laden with gold and rich furs. However, about 1650, the savages of this region commenced a wholesale massacre and all were swept away as far south as Santa Fe. Certain it is that there is a map in the archives of Paris prepared by Jesuits as early as 1792 which contains a correct topographical sketch of the Black Hills and the Bighorn Mountains."[104]

The Crow called them "Ahsahta," which means the "big horse," referring to the mountain sheep that used to roam their stony crags in abundance. Spanish and French explorers called the whole range from the British territories to the north all the way to California the Stony Mountains. When English speakers came in greater numbers they changed the name to the Rocky Mountains.

English speaking Europeans began to fill the territory like locusts and the residents of the area defended their homeland as best they could. Trappers had previously coexisted with the Tribes, often marrying Native American women, and had been accepted for many by the people for years. But miners and homesteaders started to be attacked as trespassers.

"To put a stop to these attacks, the whites in a treaty with the Indians at Ft. Laramie in 1851, agreed to give them an annuity and to leave to the Indians an area of 123,000 square miles in eastern Wyoming, Western Nebraska and Kansas, parts of the Dakotas, and much of Colorado. In return, the Indians promised not to molest the emigrants or troops along the Overland Trail."[105] But the whites broke their own treaty, not just this time, but as history shows, over and over again.

The Bozeman Trail was opened in spite of previous promises to the tribes in 1864. And in 1868, to protect the invaders who traveled upon it, military posts were established: Ft. Reno on the Powder River, Ft. Phil Kearny on Piney, and Ft. C.F. Smith on the Bighorn River in Montana. "The erection of these outposts caused the Indians to unite in an all-out attempt to repulse these encroachments upon Indian Territory.

"On a bitterly cold day, December 21, 1866—Brevet Lt. Col. Fetterman was sent to rescue an attacked wood train that was bringing vital fuel to the Fort. He was given orders to proceed directly to the train, to keep in sight of the Fort as long as possible, and not to follow the Indians beyond Lodge Train Ridge. But he was not only fiery and headstrong, but held the fighting ability of the Indian in contempt. Instead of going directly to the wood train, he swung over the hogback harum-scarum after the Indians. A few minutes later rapid firing was heard behind the ridge. Reinforcements were hurried out. They found the stripped, scalped, and mutilated bodies of the entire detachment; Col. Fetterman, Capt. Brown, Lt. Grummond, seventy-six privates and two civilians lying upon the hillside. Sixty-five pools of blood were counted within the space of an acre, showing where Indians bled fatally, although the bodies had already been retrieved, as is the Indian custom. The massacre left 115 troopers to fight off 3,000 victory-crazed warriors."[106]

Out West, Indians won many battles against U.S. forces. A notable exception was the victory of the U.S. troops in the "Wagon Box Fight" which occurred on August 2, 1867 about five miles from Ft. Phil Kearny.

Although Indians were winning battles, their war effort was doomed to failure. Many have noted the similarity between the Celtic clan culture and the Native American tribal culture. Both suffered from infighting and disunity. Both finally unified against a common enemy, but too late.

The Battle of Little Bighorn was the Indians' last stand more than Custer's. From the Indians perspective it started with the invasion of Paha Sapa (the Black Hills)—the center of the world—a place where warriors went to seek visions. But whites had found it was a place to seek gold. The "yellow fever" struck and Whites started pouring into the sacred mountains. In 1874 the U.S. army boosted their numbers to protect the white invaders. No one sought permission of the tribes and another treaty was violated by the white man.

During the "Moon of Red Cherries," Pahuska, or Long Hair as the Sioux called General Custer, led the seventh Calvary in their blue uniforms into the Black Hills. Although older tribal leaders like Red Cloud talked of peace, the people were starving and the young warriors angry. Sitting Bull, Crazy Horse, and other "wild" chiefs were invited to a council with the white man's representatives.

"If the commissioners expected to meet quietly with a few compliant chiefs and arrange an inexpensive trade, they were in for a rude surprise... The plains for miles around were covered with Sioux camps and immense herds of grazing ponies. From the Missouri River on the east to the Bighorn country on the west, all the nations of the Sioux and many of their Cheyenne and Arapaho friends had gathered there—more than twenty thousand Indians." [107]

Little-Big-Man spoke for the Sioux. According to Dee Brown in *Bury My Heart at Wounded Knee,* he was "stripped for battle and wearing two revolvers belted to his waist, 'I will kill the first chief who speaks for selling the Black Hills!' he shouted. He danced his horse across the open space between the commissioners and the chiefs."[109]

The natives had learned the white man's tactics. While most couldn't read English and never reviewed the treaty of 1868, "a goodly number knew the meaning of a certain clause in that sacred document: 'No treaty for the cession of any part of the reservation herein described...shall be of any validity or force...unless executed and signed by at least three-fourths of all the adult male Indians.' Even if the commissioners had been able to intimidate or buy off every chief present, they could not have obtained more than a few dozen signatures from those thousands of angry, well-armed warriors who were determined to keep every pinch of dust and blade of grass within their territory." [108]

The whites understood then that they'd never be able to buy the sacred Paha Sapa, so they left. Later, with the help of lawyers they acquired the mineral rights underneath the Black Hills instead. Dee tells us:

> "In the Geese Laying Moon, when the grass was tall and the horses strong, Crazy Horse broke camp and led the Oglalas and Cheyennes

north to the mouth of Tongue River, where Sitting Bull and the Hunkpapas had been living through the winter. Not long after that, Lame Deer arrived with a band of Minneconjous and asked permission to camp nearby. They had heard about all the Bluecoats marching through the Sioux hunting grounds and wanted to be near Sitting Bull's powerful band of Hunkpapas should there be any trouble." As the tribes gathered, General Custer roused his troops and they invaded the sacred hills.

On this day, June 17, 1876, Crazy Horse dreamed himself into the real world, and he showed the Sioux how to do many things they had never done before while fighting the white man's soldiers. When Crook sent his pony soldiers in mounted charges, instead of rushing forward into the fire of the carbines, the Sioux faded off to their flanks and struck weak places in their lines. Crazy Horse kept his warriors mounted and always moving from one place to another. By the time the sun was in the top of the sky he had the soldiers all mixed up in three separate fights.

In an interview given in Canada a year after the battle, Sitting Bull said that he never saw Custer, but that other Indians had seen and recognized him just before he was killed. "He did not wear his long hair as he used to wear it," Sitting Bull said. "It was short, but it was the color of the grass when the frost comes."

The tribes decided to break camp and they headed up the valley toward the Bighorn Mountains, separating along the way and taking different directions. "When the white men in the East heard of the Long Hair's defeat, they called it a massacre and went crazy with anger. Because they could not punish Sitting Bull and the war chiefs, the Great Council in Washington decided to punish the Indians they could find—those who remained on the reservations and had taken no part in the fighting." [110]

The Battle of Little Bighorn was a rousing success for the Indians but marked the end of their way of life.

MEDICINE WHEEL

Up above Dayton, Wyoming, as you follow the old cattle trail into the mountains, if you are sensitive, you may feel the pull of the Medicine Wheel.

Medicine wheels are sacred places made of stone markers placed in the shape of a circle with spokes like a wheel. This Medicine Wheel sits on the top of the world.

There are two different types of medicine wheels. The first has four spokes and is thought to be used for vision quests. It usually has a large rock in the center with more rocks marking a circle around it and four spokes which point in the four directions radiating out from the small inner circle to a larger circle.

The other type of medicine wheel may have many spokes. The Bighorn Medicine Wheel in Wyoming is of the second type.

No one knows who built the Medicine Wheel at Bighorn. Some Native Americans believe that it was made by the Plains Indians, but archaeologists believe it is much older. John A. Eddy, a solar physicist and astronomer formerly at High Altitude Observatory in Boulder, Colorado, studied the site in the 1970's. His research may have caused him to get laid off from his job.* Eddy believed and demonstrated that Bighorn Medicine Wheel may have been an ancient astronomical observatory, like Stonehenge in Great Britain. The rock structure can still accurately predict the Summer Solstice as the northern spoke is just a few degrees off of magnetic north.

No one knows the true purpose of the Medicine Wheel. To the tribal members who named this grouping of rocks, 'medicine' meant energy—"a vital power or force that was inherent in Nature itself. A person's 'medicine' was their power… Basically, the Medicine Wheel is a physical, mental, emotional and spiritual device that enables its users to attune themselves to Earth influences and forces and with the natural energies that affect their lives.

"The perimeter of the circle was marked by eight outer stones which represented powers in the universe and within man and how they could be brought into harmonious balance…

"An inner circle of eight stones encompassed the 'Source' at the centre, and these represented inner and spiritual realities.

"The remaining eight stones formed the arms of a cross, two stones being placed in each of the four cardinal directions between the inner and outer circles. These four arms represented the Four Great Paths—Love and Trust in the South, Wisdom and Knowledge in the North, Introspection and Transformation in the West, and Illumination and Clarity in the East."

---

* John A. Eddy later gained much respect and recognition for his research on the sun.

Sometimes a buffalo skull was placed at the center of the circle representing the, "mind of Wakan-Tanka—the Great Everything—and of the coming into the self of the wisdom of the creative mind of the All That Is."[111]

Bighorn Medicine Wheel is a place of pilgrimage for many Native Americans and believed by some to be "a place where spirits would go to be lifted into the Afterworld in the Milky Way…similar to the pyramids in Egypt. Inside a pyramid in Saqqara Egypt there is a book of the dead which explains the king's soul being uplifted into the sky where he becomes a star."[112]

James, Arthur, & Robert Carroll, 1914

Bob & Myrtle

CHAPTER 18

CATTLEMEN, GRANGERS, AND RANGE WARS

...There came to this country at the foot of the Big Horn mountains in the early eighties bands of intrepid pioneers who sought a new land on the last frontier where they could live and build amid beautiful surroundings. They were of a sturdy stock, descended from those who had crossed other and older frontiers to conquer the rugged hills of Kentucky, Ohio, Missouri, Iowa, and the spirit of adventure was in their veins. ~ Ruth McNally

PETER CARROLL, NICKNAMED "THE IRISH COWBOY," CARRIED A PEACEMAKER .38-40 "THAT WAS FILED DOWN SO THAT IT had a hair trigger. He carried it in a leather spring-loaded shoulder holster. Also, it shoots good and is very accurate," according to Cindy Carroll Foxworthy, the current owner of the gun.

Peter married Martha "Matty" Atherton in Iowa in 1885. Matty had spent her childhood in New York and was of Scottish heritage. Who knows how Peter convinced her to accompany him out West, but they moved to the territories around the turn of the century. One record says they moved to Montana in 1897, then moved south after five years to settle in the Sheridan area.

Peter and Matty lived in a hand-built rock cabin on a mountain above Story, Wyoming, a Shangri-La setting where the higher mountain peaks protected the little town of Story and its surroundings from harsh weather, and the streams ran clear and hopping with fish.

Matty smoked a little tobacco pipe when she stepped outside to haul wood, garden, or look at the sunset. Peter Carroll was said to have a very fine singing voice and often sang at the church services conducted by his brother Robert. Peter was prone to pulling all kinds of pranks, however, and after one episode he was asked never to sing in his brother's church again. Peter and Matty raised five children in the rugged Bighorn Mountains.[113]

Peter worked as a miner like his father, James, before him. Peter's oldest sons, Robert E. and Arthur, worked in the mines alongside him as soon as they were old enough. Robert E. quit school in the third grade to work underground and was driving mules at the age of twelve in the coal mines north of Sheridan.[114]

Coal mines were opening up all over the country and employed thousands of miners to dig the coal that fueled the booming railroad industry and new factories. The number of miles of train-tracks strewn across the U.S. had increased from around 40,000 in 1869 to over 240,000 by 1889. The railroad industry was becoming a mega-corporation.

Peter and Martha joined the community after the dust settled from the Johnson County War but before the bitter rivalry between the ranchers and the farmers (sometimes called grangers) had ended. The Grange 'movement' was now mostly over, but it continued to influence those in the Sheridan area. Many Grangers were Freemasons and like the Freemasons, they'd offered support to settlers and farm families creating bonds. Peter was a proud Freemason and a fervent believer in organized labor.

The Grange movement thrived between 1870 and 1890. The small farmers organized in order "to prevent the high freight prices for agricultural products imposed on their group by the railroad industry. The Grange advocated strong railroad regulation, government controls to prevent currency inflation, a return to the gold standard, and putting a stop to high freight rates."[115]

The railroad bosses and big cattlemen had a common enemy in the Granger and small rancher. The railroad bosses blamed the Grangers for disrupting railroad operations. The railroad bosses paid high-priced lawyers to take their case to the Supreme Court. They wanted to make the Grangers' activities unlawful. After losing their case, the railroad bosses refused to carry certain cargo or stop at some rail stations in retaliation.

The Grangers had more to fear than the railroad industry. The big cattlemen gathered at the Cheyenne Club and discussed their options on how to deal with the pesky Grangers. Most of the cattlemen were members of the Wyoming Stock Growers Association and they carried some clout. They were able to take action to discourage Grangers from homesteading on what they saw as their rangeland, needed for their ever increasing herds that they continued to bring up the trail from Texas. Ranchers and farmers erected permanent homes right smack-dab in the middle of their way, and the cattlemen were beginning to get claustrophobic. They started calling the Grangers "rustlers" to help justify their unfriendly actions.[116]

The big cattlemen said publically that rustling was the cause of their current difficulties. They'd watched rustling increase during the late eighties and early nineties with rustlers now operating boldly and openly.

In previous years, "cattle were generally tallied rather loosely, if at all, and the branding of 'Mavericks' was encouraged by the big outfits, and was universal enough to be regarded more or less as a legitimate outdoor sport, rather than a crime in the days when every rider had a rope and cinch ring at all times as a part of his standard equipment. Trail herds passing through inevitably lost a few and picked up a few animals as they crossed the country."[117]

But according to some, particularly those powerful members of the Wyoming Stock Growers Association and the exclusive Cheyenne Club, the activities of cattle-rustlers were eating away at their profits and something had to be done. There were some with connections in the south, and they hired Texas gunmen to help address the problem aggressively. The gunmen came by train and met with Association members in Cheyenne. On April 7, 1892 they started for Buffalo, Wyoming on horseback intending to clear the country of cattle rustlers once and for all, although the distinction between rustlers and Grangers was unclear.*

Grangers started to get warnings to move on years before. The warnings turned to murder as the "war on rustlers" heated up. Jim Averill and his wife, Ella Watson, were killed on the Sweetwater River in 1889 where they ran a small ranch. The cattlemen, who called themselves "Regulators," argued that Jim was a rustler and that Ella was known to their ranch hands as "Cattle Kate."

They said Ella accepted cattle in exchange for sexual favors. Stories circulated about the lonely cowhands stealing their employer's calves to pay for her company. "Apparently some of the ranchers didn't like losing cattle this way. So eager were the cattlemen to lynch the pair that they had forgotten to tie the hands of their victims and both Averill and Watson clawed at the nooses that slowly strangled them. The cattlemen left them dangling from a cottonwood tree."[118]

More likely, Jim Averill and Ella Watson were members or supporters of the Northern Wyoming Farmers and Stock Growers' Association (NWFSGA) led by Nate Champion, a small rancher brave enough to stand up to the Wyoming Stock Growers' Association and its powerful leaders.

---

* In the 1968 novel, *True Grit*, by Charles Portis, Rooster Cogburn was "hired by stock owners to terrorize . . . [those] called nesters or grangers."

Members of the NWFSGA refused the big cattlemen's suggestion that they disband and instead made known their intention to hold their own roundup in the spring of 1892.

Competing with the railroad and big cattle companies could prove more dangerous than about anything the Wild West had to offer. Jim Averill and Ella Watson were just two of many victims of the Johnson County war. Nate Champion was their main target. A well-known, local cowboy, he'd gotten on the wrong side of one of the cattle barons early on and was seen as a trouble maker. The Regulators from Texas, hired to ride against him, saw little difference between a trouble maker and an outlaw. If someone was causing them trouble, that was reason enough to end the trouble and if necessary, the troublemaker.

There were some real outlaws in the territory: Jesse James and the Hole-in-the-Wall Gang, Butch Cassidy and the Sundance Kid, and the Wild Bunch frequented this neck of the woods.

According to David F Norman, author of *Nate Champion and the Red Sash Gang*, former Sheriff of Johnson County, Frank Canton, lawman and stock detective, was himself an outlaw and wanted for murder in Texas. He was paid to look out for the members of the Wyoming Stock Growers Association.

Major Frank Wolcott, an ex-Union soldier, was one of the hired gunmen. The two Franks struck up a hatred for each other vying for control of the operation.

The train traveling from Cheyenne to Casper that April of 1892 carried them both and about fifty cattlemen and twenty-two Texas gunmen. In Casper, under the cover of night, they loaded wagons and set off by horseback, cutting the telegraph wires in Buffalo as they entered the quiet mountain town.

Nate Champion owned a small ranch with about 200 head of cattle. He was at the cabin with a few other men. The Regulators encircled his cabin before dawn, hiding in a dry creek bed to wait.

As the leader of the Northern Wyoming Farmers and Stock Growers' Association, Nate was aware he was a target and had an inkling of who his attackers were. Nate somehow managed to detail the events that led to his murder in a notebook before they killed him. A Chicago Herald reporter, Sam Clover, later published them in the paper. Clover had been hired by one of the Franks, Regulator Frank Wolcott.

From Nate Champion's Notes:

"Me and Nick was getting breakfast when the attack took place. Two men was with us—Bill Jones and another man. The old man went after water and did not come back. His friend went to see what was the matter and he did not come back. Nick started out and I told him to look out, that I thought there was someone at the stable and would not let them come back.

Nick is shot but not dead yet. He is awful sick. I must go and wait on him.

It is now about two hours since the first shot. Nick is still alive.

Boys, there is bullets coming like hail. They are shooting from the stable and river and back of the house.

Them fellows is in such shape I can't get at them. Nick is dead, he died about nine o'clock. I see a smoke down at the stable. I think they have fired it. I don't think they intend to let me get away this time. Boys, I feel pretty lonesome just now, I wish there was someone here with me so we could watch all sides at once.

I seen lots of men come out on horses on the other side of river and take after them. (Sheriff Angus was able to raise a posse of over two hundred men, all ready to take on the "Regulators.")

The Regulators took a wagon and loaded it with flammables and shoved it into the cabin.

The house is all fired. Goodbye boys, if I never see you again.

Nathan D. Champion"[119]

The Regulators left Nate's body riddled with gunshots and pinned a sign to his tattered and bloody shirt that read: "Cattle thieves, beware." Then they rode to the T.A. Ranch to get a hot meal and wait. They expected no trouble from the sheriff of Johnson County, Red Angus, believing that he had been killed as planned. While they waited, word spread from house to house about the brutal murders. Angry citizens gathered and over three hundred "Rustlers and Citizens" followed Sheriff Angus to the T.A. ranch, shotguns in hand, aiming to put the Regulators in jail.

The incident caused as much ballyhoo as the killing of the Molly Maguires. And similarly, it was accomplished with privately hired "police" who were supporting the moneyed elite, those who thought they had the right to create their own law. And like the Mollys, the Johnson County War created gaps in family bibles and left unknown spots in family histories.

Governor Barber telegraphed President Benjamin Harrison asking for help on April 12, 1892: "About sixty-one owners of live stock are reported to have made an armed expedition into Johnson County for the purpose of protecting their live stock and preventing unlawful roundups by rustlers. They are at 'T.A.' Ranch, thirteen miles from Fort McKinney, and are besieged by Sheriff and posse and by rustlers from that section of the country, said to be two or three hundred in number. The wagons of stockmen were captured and taken away from them and it is reported a battle took place yesterday, during which a number of men were killed. Great excitement prevails. Both parties are very determined and it is feared that if successful will show no mercy to the persons captured. The civil authorities are unable to prevent violence. The situation is serious and immediate assistance will probably prevent great loss of life."[120]

Conflict had reached a boiling point and local law enforcement arrested both the cattlemen and their hired guns for the death of two alleged rustlers near Casper. The Regulators were taken to Ft. McKinney by the 6th Cavalry. They were later moved to Cheyenne to await trial.[121]

"Angered that the troops and civilians had thwarted its plans, the association wired Wyoming's U.S. senator, Joseph M. Carey, on June 1, 1892 with specific demands: 'We want cool level-headed men whose sympathy is with us… Send six companies of Ninth Cavalry from [Fort] Robinson to Fort McKinney. The colored troops will have no sympathy for Texan thieves, and these are the troops we want.'"[122]

The Ninth Cavalry was an African-American cavalry regiment established July 28, 1866 in New Orleans, Louisiana. Colonel Edward Hatch was in command, and the recruits were from Louisville, Kentucky. They are often referred to as "Buffalo Soldiers." They'd proven themselves at the Battle of Beecher Island in 1868, defeating a combined force of Arapaho, Northern Cheyenne, Brulé, and Oglala Sioux braves. The motto of the Ninth Calvary was: "We Can, We Will."

The men of the Ninth Calvary were veterans of the Civil War. "The Wyoming Stock Growers' Association relied on both old animosities and a new role that black troops would assume in industrial disputes. The old animosities stemmed from long-standing tensions between white Texans and black soldiers. The new role arose from the belief that the white working class racism toward African-Americans ensured that black soldiers would harbor little sympathy for small ranchers. Subsequent events bore out that assumption." The cattlemen of the Wyoming Stock Growers' Association were correct in their assessment that bigotry could turn the situation in their favor.[123]

The hired guns were paid off and returned home to Texas. Some of the leaders among the cattlemen felt it prudent to leave for a time, too, but later returned and continued their cattle business as before. Many sought their prosecution and "an attempt was made to bring them to trial but charges were eventually dismissed."[124]

Peter Carroll brought his family into the Bighorn Mountains with knowledge and personal experience of unfairness and oppression. He'd spent his boyhood in Schuylkill, Pennsylvania. He'd lost his older brother to "foul play" related to labor organizing. Now here he was in Sheridan and the talk was still of Grangers and Ranchers and the Johnson County War. It's no wonder he felt the need to carry a peacemaker .38-40, filed down so that it had a hair trigger, in a leather spring-loaded shoulder holster. This was the Wild West, after all. Good to be prepared for anything.

FALL 1939

In September 1939, the old pioneers gathered again. Bob returned home on emergency furlough. Carrolls on both sides and Kirbys from all over came together to pay their respects to a true western woman. Mary Ella Kirby Carroll succumbed to illness and died at seventy-four in her home in Sheridan. George was devastated. Ella had been in his life for sixty of those years and it was a terrible blow, even if expected. The dreaded is never truly expected.

Harold had hurriedly dressed to attend the funeral, rushing home early from work to clean up and change. He should have graduated in June of 1938, but because of time spent recuperating from the automobile accident during his senior year, he graduated without much fanfare in June of 1939. He'd worked the last few months at P.O. Newsstand and he also managed Yale Oil Co. gas station in place of his brother Art, now in the Navy. Harold had completed the first training course of the Citizens' Military Training Camp, the basic course of instruction, during the summer of 1938 before his accident. He was still recuperating during the summer of 1939 and didn't return to complete the next CMTC course until the summer of 1940.

Harold stood beside his brothers, Bob and Art. Their mother, Myrtle and her sisters, Ednis, Jewell and Orpha, cried aloud. Uncle Granville and Aunt Alta were sad and stoic in their starched clothing. Many friends and family had gathered to offer support. The old-timers came together, some for the last time. Ella was a shining light to so many. They said goodbye to a woman and an era.

1939 was a bad year overall for the Carroll family, and the world. Granddad George's wife, Mary Ella, was dead. Harold was still recuperating from a bad car accident. Art was drafted into service after struggling to stay in college—he'd spent the 1938-1939 school year at the University of Wyoming because he lacked the money to return to Minnesota. Bob was promoted in rank in a military that faced an ever increasing likelihood of going to war, in spite of the protests of pacifist Americans.

Hitler announced Austria's "union" with Germany, in spite of the protests of Austrians. He then zeroed in on Czechoslovakia. He promised to leave western Europe alone, his focus supposedly on eastern Europe. He was betting on a strong hatred of Communism to aid his efforts. Hitler thought Britain and France would take action against him only if he aimed his weapons in their direction. So the Nazis marched east, at first.

After taking Czechoslovakia, Hitler took Poland. Next came Norway and the Netherlands; then Belgium, France....

Gathering of the Old Pioneers, August 1937.
Back Row: Alta & Granville Carroll, Ruth, Harold, Art & Myrtle Carroll, Irene & Fred Kirby. Front Row: Charley Kirby, Ira Kirby, George & Ella Carroll, Ruby & Jim Kirby

Orpha Carroll, 1927

CHAPTER 19

WINDS OF CHANGE

"Witnesses report Calamity Jane as the roughest looking human being I ever saw, and also extremely pretty. Was it red hair or raven black tresses, a degenerate alcoholic whore or angel of mercy? She definitely was at various times a wagon driver, cook, laundress, prostitute, nurse, prospector, and scout. Was she really Wild Bill Hickok's lover? She was a woman who dressed and passed as a man, a dead eye pistol shot who chewed tobacco and cracked her bull whip in Deadwood saloons. Even her birth date is disputed. It was around 1850."— Old West Tales

GRANDDAD GEORGE CARROLL AND FAMILY ARRIVED IN SHERIDAN IN 1881. SHERIDAN WAS INCORPORATED IN 1884. Wyoming became the 44th state in 1890, and it was the first state to grant women the right to vote.[125]

In 1890, Sheridan boasted a population of 281. Peter Carroll and family arrived in the territories in 1897. By 1900, there were over 1,559 Sheridan residents, and by 1910, the population was over 16,000.

Winds of change had blown through like shape-shifters. Cattlemen had driven their herds along the roaring river and altered the landscape. Grangers had staked their land claims, and they'd grown into towns. Tribes, once mighty, had been dispersed or subjugated and gathered onto reservations.

When the rail came in 1892, it brought new people, new ideas and new contraptions to the New West. Sheridan took on a new image. "In response to the 'Iron Trail' a new hotel sprang up right across the street and it boasted the first electric lights and indoor bathtubs in the county. The famous that came to the Sheridan Hotel to enjoy its new-fangled luxury included Buffalo Bill Cody and Calamity Jane, even Ernest Hemmingway stayed at the 'House of 69 gables.'"[126]

Granddad George and Grandpa Pete watched Sheridan go from a rough-and-ready cow town to a community, if not refined, offering at least indoor plumbing, electricity, restaurants, and a civic life. When they talked about the changes, though, it was not always in a welcoming way.

Granddad George took an interest in politics and became an active player in 1911. He served on the City Council in 1911 and 1912, and in 1923, was elected a member of the House of Representatives on the Democratic ticket in the 17th session of the State of Wyoming Legislature.

George was proclaimed by the Sheridan Press, dated October 3, 1911, to be "a self made man, so nearly as a man may be who starts life with a vigorous constitution, a healthy body, an independent spirit, and a mind in which the pioneering instinct is dominant, for all of these were his heritage from a long line of rugged, healthy, hearty frontiers-men and planters who for many generations have borne and honored the name of Carroll."*

Grandpa Pete served four terms in the Wyoming Legislature as a Representative for the Republican Party beginning in 1907 at the 9th Session.[127]

Peter Carroll was also an influential and recognized labor leader. He was the first president of the Sheridan Trades and Labor Council in 1907. Pete continually held office and was again president in 1927.

From the Sheridan Journal Aug. 19, 1927: "Hon. Peter Carroll, member of the state legislature and for many years a labor leader in Sheridan county has been honored by the Sheridan Trades and Labor council, he being elected president of that body. Mr. Carroll was the first president of this body when it was organized 20 years ago, and has all the time since been identified as an officer."

Peter remained a staunch, fiscally conservative Republican his whole life long, and he staunchly believed in workers' rights. Active in labor circles, he carried his union button proudly. He also carried childhood memories of Schuylkill, Pennsylvania and the never-mentioned Molly Maguires. So strongly did Peter believe in organized labor, he was known to have slept at the local union headquarters at times.

---

* Newspaper article is reproduced in Appendix D..

Like other current and former miners, Peter knew the union protected a miners' right to make decisions about how to do their work; important safety considerations like where to drill holes, place explosives, and how to secure the roof. The United Mine Workers had become one of the AFL's (American Federation of Labor) strongest unions and its members sometimes forcefully claimed their right to work without the interference of employers—employers who might be more interested in a profit than a person.

Union miners in Pennsylvania started a chain reaction across the country when they refused to let employers set their hours. In 1904 at a UMW meeting, the members rejected a proposed clause in their contract requiring "that the men stay in the mine the full eight hours." One miner quipped, "That's taking away a little freedom from a man, hain't it?" and declared, "I think a man ought to know when he is tired." Miners' resistance to supervision was so strong that they often stopped working when the boss was near.[128]

Peter fought with unrelenting tenacity for the passage of the eight-hour labor law. He complained that a lot of "busted heads and broken lives" went into creating acceptance of over-time pay. He was proud he'd helped make the eight-hour workday a reality. As economist Milton Friedman put it, "the only social obligation of a corporation is to make money. Workers still have to fight for a decent life."[129]

Pete's sons, Robert E. and Arthur, left the mines as soon as they could. (His youngest son, James, had luckily never entered the lightless life). Arthur left for the railroad; while Robert E "saved what he could of his earnings and in 1908 was married and purchased a little dry farm on the 'Bench' of Spring Willow community where he coaxed from the soil what Mother Nature could afford him. The months and years rolled by, and when the seasons drove Bob from his fields, he trampled across the countryside catching a freighter or wagon into Sheridan where he worked on week days, returning home on Friday evenings."[130] Robert was saving for a better future.

Pete's wife, Matty, died young, even for that period in 1912, of cancer at the age of forty-six. Peter lived on many more years to offer his wry advice, spiced with biting humor to those who'd listen, and to others.[131]

Robert E. achieved his dream and opened a furniture store (right up the street from Granddad George's and Granville's newly started real estate company). "World War I had come and gone. Men were returning home to readjust their lives and to fashion homes with equipment and supplies. Through his own experiences, Bob knew the discouragement of the poorly equipped family, and felt an urgency to, in some way, assist the people. In September 1919 Bob and a friend, Bob Thirwell, took inventory of their county's need and formed a partnership that was to be the beginning of Carroll's Furniture Company."[132]

Pete's and Matty's son, Arthur Carroll, met and married Granddad George's and Ella's daughter, Myrtle Carroll, on August 18, 1914. Arthur left the mines like his older brother Robert E. and took a job with the railroad. His smile is broad and he stands with an "awe, shucks" posture in front of the Burlington Route Engine train car in a faded photo.

Granddad George and son, Granville, had given up their ranging days, as had Ira and Charles Kirby, at least by horse. Charles was now an entrepreneur, traveling the country by automobile. He was still close to his little sister, Mary Ella, and frequently exchanged letters. In a postcard dated Dec. 15, 1928, he wrote in a scrawling, barely legible hand, a hand as accustomed to holding a pen as a bridle:

> *Goodnight, Dec. 15*
>
> *Dear, Sister. We came down from Matisse yesterday. Will go on Monday. It is nice and warm here we sit around with the doors open and the larks sing around of mornings. We had lots of snow almost to Matisse. It seems like another world.*
>
> *Chas. N. Kirby*

Another postcard, undated, states simply:

> *I am waiting here for Cash to come down from Denver. He is going to look around and do a little geological work for me. I am going over to see Ophilia Kirby, today. Ira has married and doing well. I like his wife well.*
>
> *Chas. N. Kirby*

The world had changed much since the Carrolls, Kirbys, and Thorns left Texas for parts northern. "Back in the day..." could still be heard around the campfire, but it was a new day. And in spite of their many differences, both Granddad George and Grandpa Pete had much in common. Both were born in 1860. Both had ventured to the wild Rocky Mountains. Both were actively involved in public life, although on opposite sides of the political fence. Both had their names on businesses in the same town and on the same block. After Ella died in 1939, both were widowers. And they shared three strapping grandsons in common: Bob, Art and Harold.

Bob & Myrtle

Myrtle & Art

Myrtle & Harold, saying goodbye at the station

DULCE ET DECORUM EST
Wilfred Owen

Bent double, like old beggars under sacks,
Knock-kneed, coughing like hags, we cursed through sludge,
Till on the haunting flares we turned our backs
And towards our distant rest began to trudge.
Men marched asleep. Many had lost their boots
But limped on, blood-shod. All went lame; all blind;
Drunk with fatigue; deaf even to the hoots
Of disappointed shells that dropped behind.

GAS! Gas! Quick, boys!— An ecstasy of fumbling,
Fitting the clumsy helmets just in time;
But someone still was yelling out and stumbling
And floundering like a man in fire or lime.—
Dim, through the misty panes and thick green light
As under a green sea, I saw him drowning.

In all my dreams, before my helpless sight,
He plunges at me, guttering, choking, drowning.

If in some smothering dreams you too could pace
Behind the wagon that we flung him in,
And watch the white eyes writhing in his face,
His hanging face, like a devil's sick of sin;
If you could hear, at every jolt, the blood
Come gargling from the froth-corrupted lungs,
Obscene as cancer, bitter as the cud
Of vile, incurable sores on innocent tongues,—
My friend, you would not tell with such high zest
To children ardent for some desperate glory,
The old Lie: Dulce et decorum est
Pro patria mori.[134]

PART IIII

Robert C. Carroll

CHAPTER 20

A BORN PARATROOPER

The skills bred into those who've faced the challenge of a harsh life in nature are skills of fortitude and flexibility (the tree bends in the wind); persistence and timing (hunting requires a lot of waiting and paying attention); generosity and competitiveness (go to a rodeo) — skills of survival passed down through generations.

Robert Collin Carroll left for military service with the West so deeply embedded in his character that its influence reached beyond the mountains and the roar and spray of the river. He'd come from a long line of people who'd worked the land and labored with their hands, and he knew the hardship and joy of the great outdoors.

Bob and his younger brothers, Arthur and Harold, had grown up hiking, climbing, staking camp, starting fires, hunting and fishing. These were activities usually done in community with others, like scouting, and Bob had reached the Eagle rank.

After graduating high school early and with honors in 1932, he worked as a brakeman for the Union Pacific Railroad alongside his engineer father. But making a career on the railroad wasn't for Bob. He left for basic training at the Citizens' Military Training Camp (CMTC) at Fort Logan, Colorado before the ink on his high school diploma was dry. The CMTC offered four weeks of hard physical work each summer with additional home study required. Successful CMTC service could help him land a commission in the military. Money was scarce and ambition hard to further during the depression, but Bob had a goal, energy, and drive in abundance, and a broad smile that opened doors for him.

In June 1933, Bob returned to Fort Logan to complete the CMTC's Red Course of Instruction. He finished the White Course of Instruction in 1934, and completed the Blue Course in 1935. In July 1935, Bob was invited to serve the CMTC as platoon leader for Company B. It was the end of the railroad for him and the beginning of a military career.

He enlisted later in the year in the United States Army and took army extension courses which he completed with scores all above the 90th percentile, and most above the 98th. Bob was twenty years old and a Private First Class. He was 6 feet 1 inch tall and had blond hair and "H. gray" eyes, according to his enlistment records.

Recommended for an Officer's Commission in the Reserve Corps in December 1936, he was appointed Second Lieutenant of Infantry. His

Robert C. Carroll, far left, at the CMTC in Ft. Logan, Colorado, about 1935

commanding officer, T.J. Sledge, Major of Infantry, rated Bob's character and efficiency as a soldier as excellent.

Bob served in the Reserve Corps during 1937 and 1938. His western upbringing had prepared him well and he was a natural in the military. He was also smart and handsome, and cut a striking figure in his dress uniform. He was highly respected by family members and others, and when he returned to Sheridan, Wyoming on leave, people would gather 'round. Friends would argue about who got to invite him to dinner, and he was generally mobbed, especially by women.

Bob and most everyone else, were not aware of the world-altering change that occurred on December 17, 1938 when Otto Hahn and Fritz Strassmann, two German physicists, discovered that the fission of certain types of atoms released incredible and unimaginable energy.

Most everyone was aware, however, of the events that led up to September 1, 1939. The French response was to send 16,000 children out of Paris, many waving a sad goodbye to parents they'd never see again. Soon the British were sending their children to the countryside, too. The reason was clear. The Nazis were coming. On September 1, 1939, World War II began when Germany invaded Poland and bombed Warsaw.

Bob was sent to Salt Lake City, Utah in January 1940. On April 24, he satisfactorily completed the required examination for AR 140-5 and was promoted to First Lieutenant of Infantry Reserve. His military career was encouraging.

The news from abroad was not encouraging, and events were brewing overseas that made it obvious to those paying attention that war with Germany was in the air.

On June 14, 1940, the Nazis marched through a defeated Paris. In July, Britain was attacked; seventy German planes bombed the docks in South Wales. By August, attacks on British ports, airfields, and industrial centers had become a daily occurrence. "On August 28, 1940, a series of four air raids on Liverpool, an area of major aircraft production, knocked the city to its knees. The raids continued nightly. London was bombed a week later."[135]

Winston Churchill sent imploring telegrams to President Roosevelt starting in May of 1940 asking for American aid. The British Royal Air Force had fewer resources than Nazi Germany. "Life expectancy of a British fighter pilot is less than 87 flying hours. Exhaustion takes such a heavy toll on the survivors that many of them routinely fall asleep as they taxi their aircraft to a stop. It is not uncommon for ground crews to remove a sleeping pilot from his plane when he returns from combat."[136] The British were getting hammered.

CMTC command, Ft. Logan, Colorado, 1935

CMTC unit, Ft. Logan, Colorado, 1935

President Roosevelt recommended American aid for Great Britain in his "fireside chat" in December. He did not commit troops, however. Roosevelt had insisted as recently as September 1940 that the U.S. was not going to war.

Hitler had already declared that American aid to the British was a violation of U.S. neutrality. He was experienced at twisting facts to his favor. Nazi plans for the invasion of France had been discovered on a plane forced down over Belgium. Hitler wisely postponed the attack on France and accused Belgium of violating neutrality. Then he sent troops to invade both countries. Hitler responded to Roosevelt in September 1940 with the following letter:

"The German people, who for one year now have been fighting against their detractors, have an untroubled conscience and know which nations before God and history are burdened with the responsibility for this gigantic struggle that is raging now. They also know who has wickedly provoked this war. They know that they themselves are fighting a just war, born of the necessity of national self-defense, out of the impossibility of solving peacefully a heavy and burdensome question of justice involving the very existence of the state and of correcting by other means a burning injustice inflicted upon us."[137]

Most Germans believed the Nazi propaganda being churned through the media and on the streets that asserted Holland and Belgium had violated their neutrality and provoked the German attacks on them.

Hitler claimed self-defense even as he attacked. His military strategy included using skydivers, paratroopers. He'd used a parachute battalion successfully, the first time, during the invasion of Norway on April 9, 1940. Hitler was certainly aware that he was on the cutting edge of a new and powerful technology, and he unleashed his newly-skilled airborne troops again in May of 1940.

Although Roosevelt was not ready publicly to commit troops, he was not going to be left unprepared or one-upped by Hitler. He insisted on having paratroops, too. "In January 1940 the Army decided to study the feasibility of air infantry and the air transport of ground troops...

"On June 25 the War Department directed the Infantry School to organize a parachute test platoon. Two officers and 49 enlisted men were selected from over 200 volunteers, and the platoon undertook a rigorous course of physical training and small-unit tactics, with classes on parachute packing and jumping. The first platoon member jumped from an aircraft on August 16. The first mass jump occurred on August 29; in September the War Department authorized constitution of the 1st Parachute Battalion, marking the Army's entry into this new form of warfare."[138]

A BORN PARATROOPER | 183

Bob may have been among the 200 hearty souls who volunteered for the parachute test platoon, but he was not yet in active service when they formed in June. Several months later, on December 15, 1940 at the age of twenty-five, Bob entered active service in the airborne division.

"The airborne division was activated at Camp Claiborne, Louisiana with recently promoted Maj. Gen. William C. Lee commanding. The airborne capability was to be provided by two glider infantry regiments (GIRs); the 327th and 401st; and one parachute infantry regiment, the 502nd, though the latter was still stationed at Fort Benning, Georgia."[139]

In May 1941, Bob was relieved from assignment with the 6th Infantry Division at Fort Leonard Wood, Missouri and assigned to the provisional parachute group at Fort Benning, Georgia.

Bob satisfactorily completed the prescribed course in ground training, parachute maintenance and packing, and made the required jumps from an airplane in flight, and was chosen and confirmed as a parachutist. Lieutenant Colonel W.C. Lee, Commander, Provisional Parachute Group, signed his promotion orders. By October 1941, Bob commanded his own parachute platoon in the 503rd Parachute Battalion.

Becoming a Paratrooper was a choice, and only selected volunteers were chosen. Men died during training. An elite group of fighters developed. This was the "new army," a tougher army. Some from the "old army" couldn't cut the mustard, or so goes the old cowboy expression.[140] N.C.O.s (Non-Commissioned Officers) came up the ranks and gradually replaced many of the "old Army" when jump training grew more intense.

The trainees wore mechanic's overalls. The obelisk and globe from the world's fair of 1939 was moved to the jump school at Ft. Benning and used for parachute jump training. Imagine the overall-clad men throwing themselves off the 250 foot structure. They hung by a single cable and dropped with their parachute. After a couple of jumps from the tower, they practiced from planes. From the roots of a tough, gritty past came unique soldiers of a new generation. The American Paratrooper was born.

Bob Carroll, Paratrooper

Robert C. Carroll, second from left

FALL 1940

They stand on the dock of a lake, the pine covered hill on the opposite side of the still water providing the backdrop for the photo shot. Bob stands between two women, his arms around them. Ruth* has short, curly blond hair and a great come-back for Bob's humorous and provocative comment. Rachels body is turned away from him as she leans possessively against him, her back arched seductively. He towers over both of them, dressed in jeans and button-up cotton shirt, smiling broadly at the camera, balancing the ladies between him, and looking lucky. The man taking the photo, Ronald, looked lucky earlier when it was his turn to get his picture taken.

The same two lovely ladies smile perkily in many shots with Bob. But he is a Carroll and so they are nature poses, and the ladies are wearing jeans. Bob, Rachel, and Ruth are pictured in the mountains horseback riding, climbing on rocks, picnicking, or just standing around the Edsel in a pine forest. Ron, odd man out, is rarely seen. Somewhere there are some lovely pictures of Ron, Rachel, and Ruth horseback riding, climbing on rocks, picnicking, or just standing around the Edsel.

Bob and Rachel often managed to slip off alone. Today, they found an oak with broad branches to shade them and thick soft grass underneath on which to recline. They didn't speak for quite some time—an uncomfortable silence for Rachel who was patiently waiting for Bob to propose. She often dropped hints when they were together, and after a long silence she spoke in a wistful tone, "Oh, wouldn't you like to just spend the rest of your life right here under this tree?"

Bob & lady friends, mid 1930's

---

*   The photos are not labeled. "Ruth" & "Rachel" are imagined names.

Bob laughed, "It'd get kind of uncomfortable when the snow comes."

Rachel slapped him playfully on the shoulder, "You know what I mean!"

Bob studied a hawk overhead. It zigzagged languidly in the clear blue sky riding on the mountain winds and eyeing the landscape below for lunch.

"A cabin right there in that clearing facing the stream, with a second story. Oh, imagine the view looking out over the valley!" Rachel leaned against him and sighed.

Bob squeezed her shoulder. He had reached a turning point in his life. The military demanded his all. There was a foul wind blowing and he felt his civic responsibility keenly. There'd be time for marriage later, he reasoned. Now was not the time.

He'd volunteered for another secret mission; he couldn't even discuss it with his friends or Rachel. He was not only an officer but a Paratrooper now. They looked at him differently; he felt different. Life had changed.

It wasn't as though his life had changed in some drastic way, more a gradual way. Jumping from airplanes was a natural next step for someone accustomed to climbing rocks barefoot and rappelling over rivers between rocky cliffs. The West bred tough folk. Granddad George could tell stories of snows so deep they'd bury the herd, of winters so long the deer starved, of drives so arduous they'd weed out the greenhorns, of rustlers so cunning and arroyos so steep... Paratroopers came from such stock.

The U.S. had reached a turning point, too. After Hitler defeated Belgium in three weeks and France in six weeks, the German people began to believe Hitler was the genius he proclaimed. And even pacifist Americans began to take his threat seriously after he bombed London.

Roosevelt had committed no troops. Many wondered if the United States would step in to defend Britain. And even if they did, could they win? Hitler had sneered that America was "corrupted by Jewish and African blood," and would be no match for the Aryan race.

Granddad George might just have warned Hitler to "watch out!" The Black cowboys he'd ridden with could handle themselves right well and he'd put them up against a Nazi anytime. It was true that many Americans came from a blend of backgrounds and cultures. But Hitler was wrong. The American military that developed was stronger because of its diversity even as it struggled with the differences. Bob looked more Aryan than Hitler, but the Scots philosophy of egalitarianism was woven deeply through his character.

Photo from Parachute Training book; Ft. Benning; GA., 1943

Arthur G. "Art" Carroll, 1941

CHAPTER 21

"90 DAY WONDERS"
It's The Cadillac of all services: Submarines. ~ Art Carroll

Arthur George Carroll graduated from Sheridan High School in June 1935 with admirable grades, winning smile, and a determined spirit. Art was tall, dark, and handsome—and a swell dancer—and he was obsessed with getting an education, and earning a degree. He'd known he was college bound since he was a young boy. With the encouragement of his family, especially his grandparents, Granddad George and Ella Carroll, he set out to achieve his dream in spite of the fact that his prospects were slim. His family had no money to contribute for tuition.

Undaunted and with entrepreneurial spirit, Art earned money for college by leasing the local Shell Service Station. Harold, his little brother, worked in the Shell station with him after school and during the summers when he wasn't at CMTC training. Harold had followed brother Bob's footsteps and was spending four weeks each summer at Ft. Logan, Colorado.

Art attended the University of Minnesota as planned, along with his cousin Robert Carroll during the 1937-1938 terms. Attending college for Art was a dream come true, and he attacked his studies with gusto. But even with the extra dollars earned from his many endeavors, he was forced instead to attend the nearby University of Wyoming his second year of college; costs were simply too high to continue to pay out-of-state tuition.

And just when he felt confident of paying the tuition for the fall term in 1940, Art was drafted into the military. Two years of college under his belt worked in his favor and meant that he received the military branch of his choice. Art chose the Navy and was enlisted in the rank of ensign. He even received his first choice of military branch — submarine service. He proudly called it "the Cadillac of all services."

Art was sent to New York where he trained on the Prairie State anchored on the Hudson River. He and the other trainees were called "90 day wonders" because they had ninety days of training there on the Hudson before they shipped out.[141]

New York City is an exciting place for a young man from small town America. Art toured the city at first with awe and then with confidence. He was with two other Navy officers when he met Mildred Holcombe. They were just three sailors alone in the Big Apple looking for companionship. One of his buddies was another westerner from Montana, David Speers. David was friends with Lorain Graider. She was also from the West, Deer Lodge, Montana.

David called her from a phone booth, "Do you know any girls?" he asked.

"As a matter of fact...," Loraine glanced around the room. There were two very bored, very female roommates at that moment looking right at her. "Yes, I do!"

Mildred was not a New Yorker either. She was from tiny Ada, Oklahoma and had graduated from Oklahoma University. She had received the prestigious Holmberg Painting Award and one year scholarship to study in Paris, France. But Paris was occupied by the Nazis in 1940, and so she was invited to attend the New York School of Fine and Applied Art, also known as the Parsons Art School. Studying in New York City was almost as exciting as studying in Paris, France.

As a young woman from small town Oklahoma in the Big Apple for the first time, helpful friends and relatives had

Arthur G. Carroll

referred Milly to a housing establishment for young women, the Three-Arts Club—"But I didn't like it," she said. Luckily, the lady next door had an apartment for rent that she did like. On the very first day she met Loraine. She was a scholarship student at Parson's Art School, too. They roomed there together at the corner of Lexington and 57th Street.

Before long they found a third roommate, Peggy Durkin. "Peg" had received a scholarship to attend NYU (New York University) and was studying for her Masters in retailing. "When she called and wanted to join the crew on 57th street, the Landlord just added another studio couch."

The three students and the three naval cadets met. Art and Milly hit it off right away. They started seeing each other regularly, on weekends and holidays, touring the city's restaurants and clubs with their friends. And quite a couple they made—tiny, perky Milly with her lustrous brown hair and sparkling eyes smiling up—way up— at tall and lanky Art, handsome and in uniform with his dark hair cropped short and his blue eyes smiling back down at her.

Milly had only a one year scholarship to Parsons. It was close to the end of that first year when her design class was interrupted and she was asked to go to the directors' office.

Milly tried not to show her concern, although it was obvious to her classmates that she was a bit flustered as she gathered her books. "They came in and got me out of class." She thought she was being sent home, but instead she was offered a scholarship to continue at Parsons for another year.

Arthur was preparing to graduate from the Naval Academy in June. Mildred expected to attend the ceremony, but her mother and brother, Floyd, surprised her at the end of the school year with a ticket home. Perhaps they had an inkling of just how serious Art's and Mildred's relationship had grown.

It was back to "Oklahoma Milly" for the summer, and a job

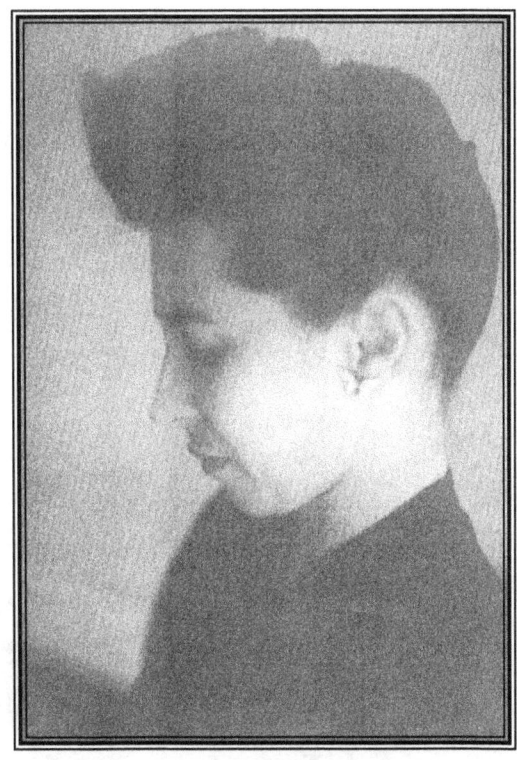

Mildred Holcombe

at Siedenbach's in Tulsa as a decorator. This meant Milly would be away from Art, New York, and all of her friends for the summer. She was sad to miss Art's graduation, but Loraine and the crew did attend the ceremony and whooped and hollered and gave Art the kind of western support to which he was accustomed.

Art expected to stay in Newport, Rhode Island for training and thought he would be close to Milly when she returned to school. Instead, he was sent to California and out to sea. He was assigned to the U.S.S. Sturgeon as navigator.

The U.S.S. Sturgeon was a "salmon-class" submarine attached to Submarine Squadron (SubRon) 6. Art and the crew departed from San Diego on November 5, 1940.

Mildred and Art wrote to each other as frequently as they could, almost every day. She decorated her letters with sometimes lovely and sometimes humorous art. Some of the letters Mildred received from Art were postmarked Pearl Harbor.

USS Illinois (BB-7) was launched on 4 October 1898. Reclassified in 1941 and renamed Prairie State, she served as a Naval Reserve Midshipmen Training School at New York during WWII.

SUMMER 1941

Harold sprayed the glass with a vinegar and water mix, and buffed the windshield crystal clear with old newspaper kept just for that purpose. He'd been working at P.O. Newsstand on Main Street since his belated graduation in June 1939. He also managed Lightning (Yale) Oil Co., the cleanest gas station in Sheridan, for his brother, Art, and for the new owner, Al Taylor, after Art shipped off to the Navy.

Harold had pretty well healed from the car wreck, although he winced whenever he thought of it and rubbed the back of his head. He had his old cockiness back as he checked air pressure and filled-er-up, whistling all the while the Cole Porter tune, "Since I Kissed My Baby Goodbye," and talking jovially with Mr. Ryan, the customer in a black Ford.

A new 1941 Cadillac surprised him from his reverie with a sharp honk of its melodic horn. Harold collected money from Mr. Ryan and waved the shiny car over.

It wasn't until just this summer that he'd become aware of a class system in Sheridan. In high school, smarts and sports and good looks leveled the playing field. But here in the adult world, there were those who owned businesses, and those who did the bidding of owners. Harold dreamed of going to college and becoming a lawyer and having the power to right injustice. Harold was smart, funny, and good natured, but he had a short fuse and was fervently proud.

Bob Burns rolled down the window of his fancy jalopy and sneered by way of a greeting. He wore new duds in the style of the season, and he was fussily dressed for business, or courting, or church, or trouble. Harold doubted that Bob Burns had ever done an honest day's work in his life. Today wasn't Sunday. That left trouble of one form or another.

"How's life treatin' you, Carroll?" he asked insincerely, looking pointedly at Harold's work-overalls and the station sign, Yale Oil Co.

"Fill-er-up?" Harold asked impatiently.

"There are some nasty spots on this windshield," Burns gestured with his hand toward bird droppings on the window. "I bet you're the master at getting these off. You always were good at spot removal," Bob snickered.

The great Burns, home from college to lord it over the townies. "...should have stayed in Denver," Harold thought, or may have mumbled. He kept a cool exterior but he was getting hot under the collar.

Good at spot removal? What had he meant by that? "I'll give him spot removal," Harold mumbled, "and he's the spot!" Harold squirted cleaner on the windshield and started polishing. Then he squirted a bit more, and polished harder, and squirted a bit more—right there in the corner near the still open driver's side window where

Lowry Field, Colorado, Air Corp Training (ACT) School, Photo Technicians, H back row third from left

Bob Burns burr cut was sticking slightly out while he cleaned his teeth in the side mirror…

"You Son of a Bitch!" Burns screamed, wiping fitfully at his damp dome.

Harold demanded he get out of the car, "I cannot allow you to insult my mother."

The fight that ensued cost Burns his eye tooth and Harold his job. He enlisted in the Army July 1, 1941 in Ft. Warren, Wyoming as a private.

Shortly after boot camp, Harold disappeared— gone for all practical purposes because everyone thereafter called him "H." That is, everybody except Bob and Art, of course, who after the fight just called him stupid. He didn't hear too much from them, luckily, as they were both away at their duties. Lt. Bob was a paratroop officer now, and insisted that H become an officer, too, so they could go to the officers' Club together. Lt. Art was in the U.S. Navy preparing, but in no way prepared for December 7, 1941. George's and Peter's grandsons were now military men.

H Carroll

CHAPTER 22

PHOTO LAB NON-COM

It was a very basic cardboard box camera with a simple meniscus lens that took 2¼-inch square pictures on 117 rollfilm. With its simple controls and initial price of $1, it was intended to be a camera that anyone could afford and use. ~ Wikipedia's description of the Brownie.

Boys going to war look completely different than returning veterans. Not wanting to be outdone by his daring older brothers, Harold James Carroll enlisted at Ft. Warren, Wyoming in July 1941. His picture, taken just before leaving home, shows him blaring white dressed in khaki, teeth blazing, sparkling eyes—a patriot in waiting. Although shorter, younger, and less experienced than Bob, he resembled his older brother in looks and charm.

Harold was twenty-one years old and experienced in the ways that nature and a small town educates, yet naive as to the ways of the world he was about to enter. He eagerly left for Ft. Leavenworth, Kansas as a trainee. Leavenworth seared in the late summer sun. The rolling hills of wheat, instead of soaring snow-capped mountain peaks, made him feel far from home, but he was following in his brother Bob's admired footsteps, like he had countless times before, and he followed with gusto. He felt just a touch of trepidation he would never admit.

In September 1941, across the Atlantic Ocean, awaited a reality more harsh than Bob, Art, or Harold could then have imagined—the Nazis ordered all Jews to wear yellow stars and the gas chambers at Auschwitz were used for the first time to kill people.

Harold soon left for Lowry Field, Colorado, Air Corp Training (ACT)

School where he became a photo lab non-com (non-commissioned officer, or NCO). The technical training included aerial photography and the school at Lowry Field had trained image interpreters since the beginning of the war. Harold's talent and love of amateur photography was professionalized and put to good use by the Army, as was his brazen daring. He was taught to parachute from planes while carrying giant, aerial cameras, and then to interpret the images.

In addition to military photographs, Harold took pictures of his comrades and his surroundings on his old, but tenderly-treated Brownie. His pictures depict the innocence of young men having never yet seen the brutality of war. That would soon change.

Lowry Field, Colorado, Air Corp Training (ACT) School, H front & center

The U.S.S. Sturgeon left Hawaii on November 10, 1941 for the Philippines. Lt. Art Carroll immediately struck up a friendship with another young officer, Chester Nimitz, Jr. They'd been sent to patrol the area between the Pescadores Islands and Formosa. Art and Chester and the rest of the crew moored in Mariveles Bay until November 22, 1941 when the Sturgeon continued on to Manila Bay. There, the sailors spent time ashore sleeping in barracks on cots on the solid ground instead of in berths tossing and turning on the rolling sea.[142]

December 7, 1941 the skies above Pearl Harbor on Oahu, Hawaii darkened with aircraft, red circles decorating their wings. The Japanese struck unexpectedly while many onlookers just stood and stared. There were an estimated 2,333 American lives lost that morning and another 1,139 or more wounded. The U.S. Navy was gutted with five battleships utterly destroyed, as well as one hundred and eighty-eight aircraft, one minelayer, and three destroyers. The public was not informed as to just how extensive the damage was to U.S. naval defenses.

Commander in Chief of the Pacific Fleet, Admiral Chester W. Nimitz, Chester's father, spoke forcefully after the surprise attack, "We must not again underestimate the Japanese…It is the function of the Navy to carry the war to the enemy so that it will not be fought on U.S. soil."[143] Chester and Arthur Carroll were leading the way in the U.S.S. Sturgeon.

The prospect of war with Japan was now seriously considered; strategies had to be argued and plans drawn up. The so-called Orange Plan advised that the U.S. should proceed through Micronesia and seize strategic islands while heading westward toward the Philippines. This plan had its drawbacks. It required U.S. troops be able to "hold Manila Bay for up to six months while the Pacific Fleet and its accompanying ground forces conducted an island-by-island advance."[144]

The U.S. was forced, however, to adopt a much more defensive

Harold J. Carroll

strategy in the Pacific. The crew of the Sturgeon patrolled cautiously if aggressively, sometimes finding more than they could deal with.

"The submarine found a convoy of five merchantmen accompanied by a cruiser and several destroyers on 18 December. As she came to periscope depth within attack range of the cruiser, she was sighted by one of the escorts approximately 250 yards away. She started going deep but had only reached a depth of 65 feet when the first depth charge exploded, breaking numerous light bulbs but causing no serious damage. Sturgeon began silent running and evaded the escorts."[145]

The Japanese attacked Wake Island on December 23, 1941. On December 15, 1942 the British surrendered at Hong Kong. Then British North Borneo surrendered, and Singapore on February 15, 1942. "With the fall of the Philippines on 6 May, the Japanese seriously threatened to achieve their goal of dominating the Pacific basin."[146]

Late one night, alarms awakened the crew. They'd been warned that a large convoy was headed their way in Makassar Strait. "A few minutes later, her sonar picked up the pings of ships dead astern. She submerged and fired four torpedoes at a large ship, with two explosions following. The submarine was then subjected to a two and one-half hour depth charge attack by two destroyers which caused no damage."[147] They'd survived another close call.

The U.S.S. Sturgeon was not always on the run. In January, 1942, during an encounter with a Japanese transport ship, she "fired a spread from her forward tubes which resulted in a large explosion on the transport, and her screws stopped turning. No post-war record of a sinking could be found, but the transport was believed damaged. Three days later, she made two hits on a tanker."[148]

H Carroll, camera in hand.
Photo from canister developed in 2006.

While Art navigated the U.S.S. Sturgeon through the dangerous waters of the Pacific, his little brother, Harold, was learning the ropes, so to speak, in the Air Corps. H was soon to follow in Bob's footsteps by instructing soldiers how to properly throw themselves out of airplanes.

Photo collage created by H Carroll at Lowry Field, Colorado

WINTER 1941

"Cribbage!" Art declared forcefully but quietly. It was Christmas Day and Manila was being bombed by the Japanese. 1st Lieutenant Arthur Carroll, Navigator for the U.S.S. Sturgeon, now under Manila Bay, lay down his cards.

"Shhhhh!" from a couple crew members reminded him that he hadn't been quiet enough. The sound of sonar, louder than anyone might have imagined prior to spending time on a submarine, usually squawked incessantly. It was silent now as they operated in passive mode, unable to get information about other vessels without revealing their position. They had to be careful. A skilled sonar operator could tell how fast a submarine was going, the number of propellers, and more just by listening to the sounds.

Art and the rest of the Sturgeon crew stationed in Manila, had been trying to follow orders, keep busy, and make themselves comfortable in a place of tall palms and tropical breezes.

All of Art's belongings, the few that he had, were stowed topside in one of the barracks that sat like giant, shiny, silver roly-poly's lined up in rows on the lush, green island. The crew's belongings were now being systematically destroyed, blown to shreds by the Japanese.

The sailors had no time to gather pictures, or mementos, or even socks, or cigarettes before the diving bells rang and they dove for safety to the bottom of the bay— a place that might have been lovely if there wasn't this war going on. If it were peaceful and if Mildred were near... Art didn't have much time to ponder such fantasies. He'd lost his Sigma Nu pin under the rubble that was left of Manila after the bombing. That college pin had meant a lot to him and its loss made him even more determined to pursue his education when this godforsaken conflict ended. Later, when the waters were calm, he wrote to Mildred, "I need a new pin!"

Alert and dejected, the crew occupied themselves as best they could in the close quarters of the submarine that Christmas Day of 1941. The commander announced graciously, "We might as well enjoy Christmas dinner down here." And it was served to the best of cook's ability: salt pork dressed with canned pineapple and brown sugar, green beans with bacon, and his piece-de-resistance, apple compote ala sort of whipped cream, or when pressed, warm apple mush with somewhat soured milk.

"You lucky bastard!" Nimitz exclaimed.

"I assure you both of my parents are happily married," Art shot back.

"Just not to each other, huh?" quipped Fred, a skinny sailor with a big smile.

Someone began to whistle a Benny Goodman tune.

"I was there!" Fred reminisced. "I was there when Benny Goodman and his orchestra gave the first Carnegie Hall jazz concert!"

"No!"

"I don't believe it!"

"It's true!" he insisted. "There was Count Basie and Duke Ellington, and…"

"Shhhhh!" someone reminded again. They had to take care to be silent. Even playing cribbage by candle-light on the bottom of the bay carried its risks.

"All That Meat and No Potatoes" was playing loudly on the radio. Mildred couldn't help but hum along with Fats Waller's jazzy tune while she worked on her latest sketch. Milly was in the middle of her second year at Parsons School of Design and had been offered a job working at the studio of Joseph B. Platt, a prominent man in the design industry who also served on the Board of Directors of Parsons. He mentored the budding talent of petite, pretty, and sweet Mildred from Oklahoma in a protective and fatherly way.

Platt's studio had earned a name in New York. Joe Platt and his crew designed sets for Broadway shows like Gone With the Wind and created and designed building interiors and window displays for Macy's and Gimbal's and others. And Platt's Studios also designed private spaces for the very rich.

Mildred drew up designs and her staff carried out her ideas. One of Platt's clients, Stark Ribbon Company, required all of their show rooms to be redone annually. Mildred enjoyed designing fabrics with intricate patterns to be hand-woven.

New York elite allowed the use of their apartments for advertising. They'd redecorate and have a party, a very public party with media invited. Mildred was on a team that designed Helena Rubenstein's apartment. Included in Mildred's up-and-coming crowd were Marvin Monk who also worked for Platt's Studio. He later became the Art Director for Lord & Taylor.

Without much coercing, the three girls, Milly, Peg and Loraine accepted another roommate; a non-westerner joined their N.Y. city pad. Perky Pauline Jackson was from New Jersey. She wanted to be a freelance artist.[149]

So here were Mildred, Loraine, Peg, and Pauline, and they needed more room as space was scarce and the bathroom crowded. They went looking for a place to move, sometimes dragging along a handsome sailor or two. A perfect, but short-lived solution for the four ladies all interested in pursuing their education in the heart of Manhattan was adjoining apartments on 1st Avenue. Moving took the assistance of perhaps half of the U.S. Navy!

The U.S.S. Sturgeon left from Hawaii before the attack on December 7th. Mildred exhaled a conflicted sigh of relief knowing the Sturgeon had left for the South Pacific with Art safely aboard. She sat now crossed legged in the new apartment amid the clamor of people celebrating the holidays. She'd saved her last letter from Art postmarked from Pearl Harbor, Hawaii, and was rereading it again when Loraine sat down beside her and put an arm around narrow shoulders.

"He'll be OK," Loraine wanted to say, but she just squeezed Milly's shoulders instead.

In another part of the world, 1st Lieutenant Robert C. Carroll headed for the privacy of his quarters. Once there, Bob turned the envelope over and over in his hands, noting the scent of lavender and the delicate cursive script spelling his name. He had a bad feeling. Finally, he tore it open and read the single sheet of paper. Rachel's words were brief. She understood that he was devoted to the military, and so unable to devote himself to a life with her. She'd met someone...

He cursed himself! There was so much he'd left unsaid, especially about their future together. She'd dropped hints. "Oh, wouldn't you like to just spend the rest of your life right here under this tree?" There was nothing he wanted more. And yet, he'd never responded. As he tossed the letter and envelope in the small trash receptacle in his quarters, his eyes were shining with moisture.

Milly at Work

"REMEMBER PEARL HARBOR"
Song lyrics, author unknown:

Let's remember Pearl Harbor as we go to meet the foe.
Let's remember Pearl Harbor as we did the Alamo.
We will always remember how they died for liberty.
Let's remember Pearl Harbor and go on to victory.

Paraskiiers Hit the Slopes

CHAPTER 23

SKYMEN & SUBMARINE WARRIORS

They're only six months old. War babies, you might call them. But when hordes of them come tumbling out of the clouds in the frozen north next winter to spread death and terror behind enemy lines, they won't be babies any longer. They'll be one of Uncle Sam's surprise packages for Messsrs. Hitler, Hirohito, et al—the wild, tough, gritty, intrepid hellions of the United States 'paraski' troops. ~ Don Eddy, International News.

THE SPRING OF 1942 WAS FAST PACED FOR BOB, ART, AND HAROLD. H LEFT BY TRAIN FOR BOWLING FIELD, NORTH Carolina in February 1942. He joined the 1st Mapping Group, 19th Photo Squad, where his knowledge of photography would be taken to new heights. He would be doing something that few had before attempted: leaping from a plane with a camera strapped to his body.

Aerial photography had been used successfully during WWI, before the advent of paratrooping. "In 1940, the GHQ Air Force had one photo squadron with six flights scattered throughout the United States, each flight equipped with a single F-2A, a modified civilian Beechcraft twin-engine transport, and a few reconnaissance squadrons equipped largely with obsolete B-18 bombers...

The twin-boom P-38 Lightning became the F-4 and F-5 reconnaissance aircraft when cameras replaced its guns, and they earned a reputation as the reconnaissance workhorse in almost every theater during World War II. At one time or another practically every type of tactical aircraft was modified to carry one or more cameras."[150]

In some circumstances, the cameraman would use the both the gun and the camera, hopefully without confusing the two.

H was promoted to Photo Technician Sergeant; he was gaining skill and confidence and moving up in rank, aiming for those officer's insignia so that he and Bob could go to the officer's club together.

Bob was promoted from 1st Lieutenant to Captain in February of 1942, and assigned to the Battalion Staff of the 503rd Parachute Infantry.

In March of 1942, they were testing the slopes of the Wasatch Mountains of northern Utah, or perhaps the slopes were testing them. Bob was an expert skier and was just as "at home" on skis as he was in freezing, snowy Rocky Mountains.

These "SkyMen" were expected to "write their name in glory."

When the photographers for a New York paper aimed their flashes at Bob, below far right, he smiled just enough to show his dimple and stood to his full 6'2 inches. (He'd grown a bit since enlisting, or his stretching was paying off!)

Paraskiiers Hit the Slopes, Bob Carroll far right

INTERNATIONAL NEWS, MARCH 5TH, 1942

Until recently, parachute troops and ski troops were two different kinds of troops. As a general thing, they still are. But, as we prepared to fight on all the fronts of the world, including the Arctic, somebody in Washington got a bright idea.

'We should have ski troops who can parachute from airplanes,' he must have said to himself. 'They can be dropped into snowbound country far behind enemy lines, for sabotage or surprise attacks or both.

A call for volunteers went through certain sections of the Army, and thus the 'paraskis' were born. They have grown from a few squads to many thousands. Their exact number and the many phases of their training and battle duties are closely guarded military secrets, but it's not secret that our strategist expects the paraskis to write their names in glory before this war is over.

I can testify to their daring and hardihood. By special permission, I have just visited one contingent at a training ground on the snow-locked crest of the Wasatch Mountains of northern Utah. Like most roving reporters, I had considered myself a fairly husky specimen; now, alas, I know I'm just an old softy. I have chilblains, a bilious-looking toe, and a face that feels like something out of the tool shed.

And I have something more: a profound respect for these high-flying daredevils...A paraski trooper, in his natural element, looks like a ghost; he is virtually invisible. He wears a loose-fitting white coverall with a parka hood. His skis are white, as are his shoes and gloves, and under battle conditions he will whiten his face, pack, and rifle. Once, in a dense snowstorm, a column passed me in a single file not ten yards away. I could hear the swish of their skis and an occasional clink of their equipment, but I couldn't see them at all.

They are, however, infinitely more deadly than most ghosts. They carry Garand rifles and an assortment of trick sidearms, some reminiscent of old Indian fighters...With paraski troops, more than with ordinary Army parachutists, the rapid-fire exit system is vital, since the men must land within reasonable distance of their skis and packs. Elapsed time in wallowing through 10 feet of snow might be fatal.[151]

The photos are crisp and clear black & whites. In one photo the men pose in formation on the slopes, goggles up, grins on. In another they are making coffee, (or trying to.) Bob's job was to not only train, but maintain authority and control over these twenty or so guys pictured having a great time on skis in the snow!

Paraskiiers Hit the Slopes, Bob Carroll far right

Paraskiiers with Reporter, Bob Carroll second from left

While Bob navigated the Utah slopes, middle-brother Art navigated the Sturgeon in the South Pacific. The U.S. had taken Admiral Nimitz's advice and had carried the war to the enemy.

"On the morning of 8 February, Sturgeon found herself on the track of an enemy invasion fleet headed toward Makassar City. She submerged to avoid detection by several destroyers and a cruiser, as they passed overhead, but was able to report the movement of the convoy to Command."[152]

Reporting information was crucial. Much of the U.S. Navy's Pacific fleet was out of commission after the attack on Pearl Harbor. Submarine warfare became one of the best and most relied upon weapons the U.S. possessed. Although they made up only about two percent of the U.S. Navy, submarines destroyed over thirty percent of the Japanese naval forces. The Sturgeon and crew played their part.

Naval records show that on March 30, 1944 "she sank the cargo ship Choko Maru. On 3 April, one of her torpedoes caught a 750-ton frigate directly under the bridge, and she was officially listed as probably sunk. She then fired three torpedoes at a merchantman but missed. With one torpedo remaining in the bow tubes, she fired and hit the target abreast the foremast."[153]

Back home, later in April, Japanese-Americans, entire families, mothers with babies and grandfathers with canes, packed their belongings and awaited transport to relocation centers. Most would be locked up for the remainder of the war. Many had been born in the U.S., but "America the free" was selective with freedom. There were those who believed that enemies lurked within U.S. borders. And an enemy certainly did thrive within the U.S.—that enemy was fear.

Fear of the Japanese was well founded by the crew of the Sturgeon and others located in the Pacific. What must it have been like, within the cramped quarters of the submarine, trying to navigate the crew to safer waters, or worse navigating straight into battle? "On 28 April, the submarine sailed for Australia. However, she interrupted her voyage on the night of 30 April in an attempt to rescue some Royal Air Force personnel reported on an island at the entrance of Cilacap Harbor. A landing party under Lieutenant Chester W. Nimitz, Jr. entered the cove and examined it by searchlight but found only a deserted lean-to."[154] The former occupants had disappeared!

On June 25, the U.S.S. Sturgeon was on patrol in an area west of Manila when she caught up with a nine-ship convoy before daylight. She fired three torpedoes at the largest ship and they heard explosions. Then the Japanese retaliated. "After some 21 depth charges were dropped by the escorts, she managed to escape with only a few gauges broken."[155]

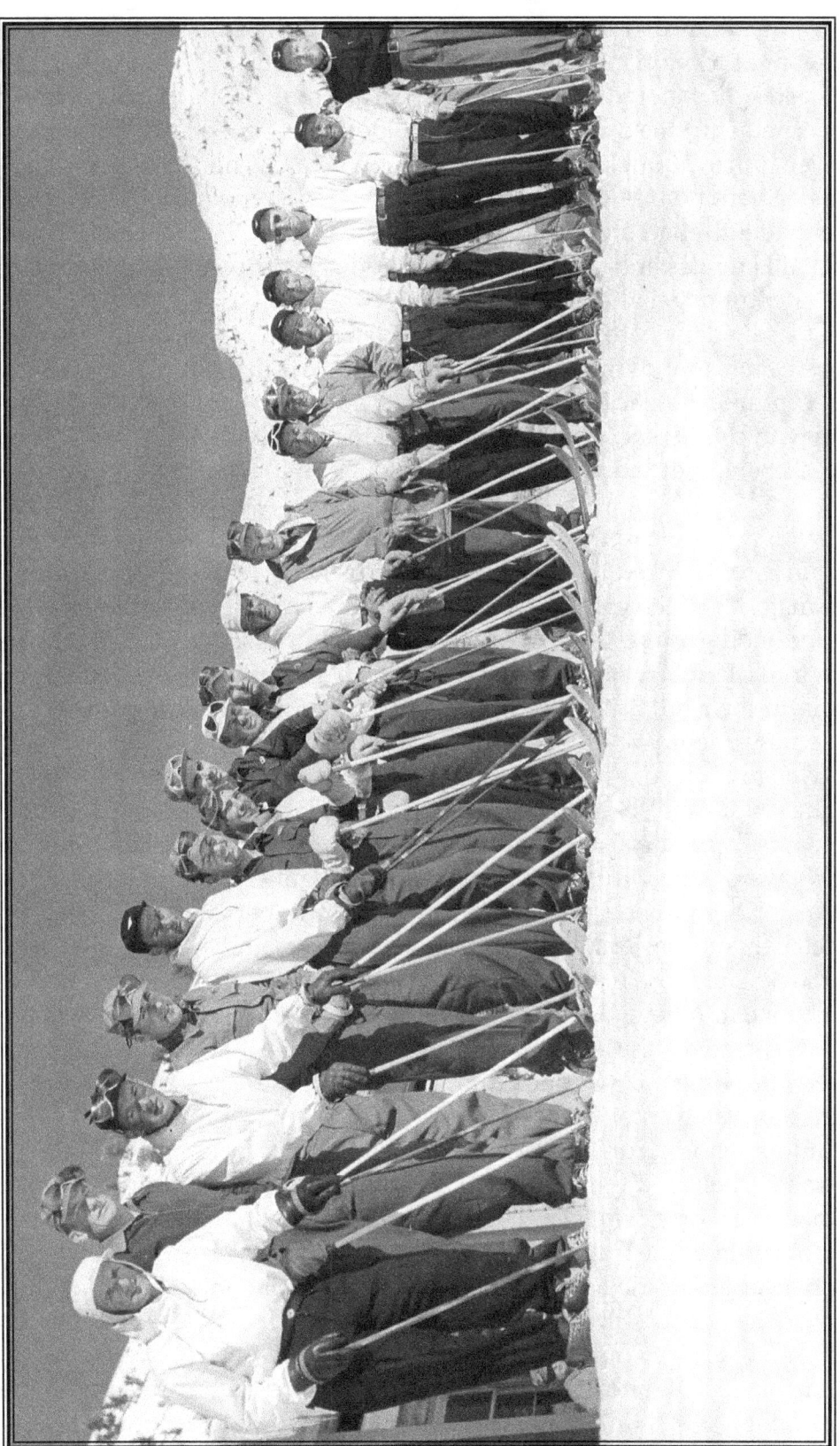

Posing Paraskiiers, Bob center, March 1942

On June 30, near Luzon Island in the Philippines, the skipper of the Sturgeon, Lt. Commander Wright, wrote in his log: "Patrolling northeast of Bojeador as before. Dove at dawn, surfaced at dusk. At 2216 sighted a darkened ship to southwest. At first, due to bearing on which sighted, believed him to be on northerly course, but after a few minutes observation it was evident he was on a westerly course, and going at high speed. He quite evidently had stood out of Babuyan Channel, headed for Hainan. Put on all engines and worked up to full power, proceeding to westward in an attempt to get ahead of him."[156]

Just a few days before, on the nearby Island of New Britain, Japanese troops began rounding up more than 1000 Australian soldiers who had been captured defending New Guinea. The Troops rousted the POWs, along with civilian aid workers, including the Salvation Army band. They were all being herded to Rabaul Harbour to be loaded onto a transport ship.

The Japanese woke the Australians at about two o'clock one morning. Captain Peter H. Brown described what happened in a post-war interview: "The majority of the camp had no suspicion what so ever of the fact that we were going to be sent away. We were all aroused and orders were given to pack up and prepare to be searched. The officers were all told to remain in their hut and were not to leave under any circumstances. The whole movement was carried out by Japanese naval personnel and was very well organized, as were the majority of naval operations.

"The sailors and their officers arrived with placards labeled 'Civilians,' 'Sergeants,' 'privates,' etc., and in quite an efficient manner different groups were called forward with their belongings and searched…The vast majority marched out of the camp but one person was carried on a stretcher and several limped out on crutches and were assisted by comrades. The morale was terrific as the majority of people were glad of any change to relieve the monotony and secretly thought we were being moved in view of the Allied attempts to recapture Rabaul, and that the movement would be to a neighbouring island there was a very strong rumour to that effect."[157]

Rumors tend to run high where information is lacking. Some of the Australians believed they would all be sent to Japan. Others insisted that the men over forty-five were being returned to Australia. Eight hundred and forty-five Australian POW's and two hundred and six civilian prisoners boarded the Montevideo Maru bound for somewhere.

From the log of Lt. Commander Wright, U.S.S. Sturgeon: "For an hour and a half we couldn't make a nickel. This fellow was really going, making at least 17 knots, and probably a bit more, as he appeared to be zig-zagging. At this time it looked a bit hopeless, but determined to hang on in the hope he would slow or change course toward us. His range at this time was estimated at around 18,000 yards. Sure enough, about midnight he slowed to about 12 knots. After that it was easy."[158]

The Montevideo Maru had been a passenger and cargo vessel, and normally shuttled between Asia and South America prior to WWII. The ship was used by the Japanese to carry troops and cargo, and now prisoners of war. She had no markings to indicate the presence of POWs or the Salvation Army band onboard. They left Rabaul Harbor to meet with two Japanese destroyers that were to escort the ship to Hainan, China. But the destroyers were late. The Montevideo Maru slowed her course.

The Sturgeon fired four torpedoes the morning of July 1, 1942. The Montevideo Maru went down. There were only eighteen reported survivors, all Japanese.

Some believe the Australians were murdered by the Japanese troops or taken away to work as slaves. There are even conspiracy theories surrounding the Montevideo Maru. In an interview on "ABC Online 7.30 Report," a former sailor said that some Australians did survive and apparently ended up as prisoners in Japan.

The official reports say all the Australians drowned. However, one witness was convinced that they did not all die. An interview of this witness was aired on Australian radio in 2007:

> **YOSHIAKI YAMAJI:** I was told by an official of the company that the POWs were saved by a destroyer and taken back to Kobe. They came back earlier than I did and worked at Kobe. I do not believe all of them died.
>
> **MARK SIMKIN:** This startling revelation is supported by at least one naval historian. Hisashi Noma has investigated the case and believes the destroyer scheduled to rendezvous with the 'Montevideo Maru' saved the Australians.
>
> **HISASHI NOMA, NAVAL HISTORIAN:** The sea was calm, not rough sea. And the season was the south sea in July or June-July. No reason why they could not keep themselves alive long time.
>
> **MARK SIMKIN:** Albert Speer is another who is pursuing the truth. He served in New Guinea during the war and says this 1942 newspaper photo backs up Yamaji-san's story. It allegedly shows Australian POWs at work near Tokyo, POWs who had supposedly died months earlier on the 'Montevideo Maru'.
>
> **ALBERT SPEER:** The Commanding officer of this gunnery battalion recognized his two former soldiers and he said, "This is my men"…
>
> **MARK SIMKIN:** And yet the uncertainty continues.
>
> **PROFESSOR HANK NELSON:** The people in the photograph are wearing American uniforms, they are identified in the captions below the photograph as being British and American prisoners. There's no mention of Australians.

MARK SIMKIN: Hank Nelson is not convinced by Yoshiaki Yamaji's story, saying the claims about Australain survivors making it to Kobe are hearsay.

PROFESSOR HANK NELSON; I think that most of the Australians would have gone down with the 'Montevideo Maru', that a few of them would have surfaced and that they would have been sadly abandoned to their fate by that disappearing Japanese crew. It is unlikely that a Japanese destroyer would have stopped and remained still and picked up prisoners when they know that there are American submarines about.[159]

Many continue to believe that Australian POWs survived the destruction of the Montevideo Maru. "The question of the uniforms seems quite ridiculous. What uniforms would men who had left the tropics of Rabaul and crossed the Equator when the Montevideo Maru was sunk have been wearing at the time of the sinking? What pieces of their uniform would have survived the sinking? Historians also argue that Japanese destroyers wouldn't have stopped to pick up the Allied men in the water due to the submarine being in the area. There is at least one precedent in the Philippines where in fact a Japanese ship did stop and pick up Allied survivors."[160]

From the log of Lt. Commander Wright, skipper of the U.S.S. Sturgeon: "At 0225 fired four torpedo spread, range 4,000 yards, from after tubes. At 0229 heard and observed explosion about 75-100 ft. abst stack. At 0240 observed ship sink stern first. 0250 surfaced, proceeded to eastward, completing battery charge. Ship believed to be Rio de Janerio Maru, or very similar type, although it is possible it was a larger ship, he was a big one. A few lights were observed on deck just after the explosion, but there was apparently no power available, and his bow was well up in the air in six minutes. Dove at dawn, no further contacts."[161]

The "fog of war" is a phenomenon affecting naval as well as land troops. Cell phones, the Internet, and Facebook didn't exist for the crew of the U.S.S. Sturgeon. The events of 1 July, 1942 regarding the Japanese ship, Montevideo Maru, remain a mystery even today. Most believe that she was destroyed by the U.S.S. Sturgeon and that more than 1000 POWs, secured in the bowels of the unmarked, unescorted ship, were killed. Considered to be the largest maritime disaster in Australian history, the Japanese did not report the loss of the ship or the Australian prisoners.

Art and Chester never knew about the tragedy as the facts did not surface in their lifetimes.[162]

SUMMER 1942

Most everyone was surprised when H managed to get himself stationed at Fort Benning, Georgia, as that required some real manuvering on his part.

"Harold had a hard time getting into the same outfit as his older brother Bob. He had to finagle and push and wheedle and everything else because he wanted to be there. But even Bob discouraged it and said he didn't think it was a good idea. Harold was so persistent they finally let him in," Mildred remembered. "He was just as determined as can be."

Not only was H tenacious, but he could tell a story better than anyone. "He had a wonderful style and sense of humor!" according to Mildred. The events of one story H told Milly occurred while he was training at Fort Benning, Georgia. Mildred recounted the story:

> "These three paratroopers decided they didn't have anything else to do. So one Sunday they went to visit the Holiness Church, an evangelical church, near where they were stationed.
>
> "There they were, seated in the back pews, trying to understand the service. The minister was dancing back and forth and singing and chanting and going on and on, and a woman fainted up in front.
>
> "Well, these three paratroopers rushed down the aisle to help her, but the minister didn't break a beat, he just kept right on going and bellowed to the paratroopers, pointing dramatically at the woman, 'Leave her lay where Jesus flung her!'
>
> "The three boys looked at each other, made an about face, and turned and went straight back up the aisle and out of the church."
>
> And that became the saying we used with the children when they were little and having tantrums: "Leave her lay where Jesus flung her!"

In the neighboring state of North Carolina, Bob nursed a beer and engaged in witty conversation. Lt. Col. Robert Collin Carroll and Lt. Col. Phillip S. Gage postured in the officers' club in Pinehurst with their drinks raised. Bob and Phil were great friends. Across from them were Phil's fiancée, Susan, a regular on his arm, and her sister, Patricia, visiting North Carolina for the first time.

Bob couldn't keep his eyes off of Patricia. He'd sworn off women, or so he said. He was still stinging from Rachel, knowing she was married now to someone else, and with a kid already in tow, and another on the way. But Bob didn't blame her. She simply couldn't count on a future with a military man. This blasted war! Looking at Pat erased everything else from Bob's mind. He was captivated and entranced.

The officers were here for much deserved R & R after months of intense training.

Susan had planned to do some catching up with her sister during the visit. But Bob and Pat spent the entire evening dancing and talking as if in their own little world together, even when Phil and Susan were at the same table.

Phil was surprised at Bob's response to Pat, "He's clearly smitten!" Phil and Susan talked about it while they sat alone together watching the two on the dance floor.

"I thought he was married to the military!" Susan laughed.

"He is!" Phil shook his head in wonder.

Susan leaned toward him conspiratorially, "I think she's falling for him."

"Well, you know, Pat could be just the one to give him something to look forward to when this stinking war is over."

Phil and Susan both decided the two should get together. They just didn't know exactly what they could do, except hold hands and watch it happen.[163]

In another part of the world, Hitler decided to get rid of the Jews. His only difficulty was figuring out exactly how to accomplish their elimination.

After taking over Poland there were even more of them. All ideas were on the table, including deporting all Jews to Madagascar, or confining them in Poland, both unrealistic notions. He'd already determined that only about forty percent could be used for forced labor so the only effective solution was the final solution—kill them.

Heinrich Himmler, military commander and leading member of the Nazi party, used his specially trained soldiers from the Schutz Staffeinel, (SS), to implement the extermination program. "At first the victims were shot, but with a high proportion of those involved in the killings suffering from nervous breakdowns, a more impersonal method was developed," hence the invention of the gas chamber."[164]

As spring dawned in 1942, over 500,000 Jews in Poland and Russia died at the hands of the Schutz Staffeinel (SS).

from the 1942 503rd Parachute Battalion Historical and Pictoral Review

H Carroll

CHAPTER 24

LOVE AND WAR

The term of the day was "khaki wacky" and the two beautiful people fell in love in a weekend, and next stop was marriage—right away, as Dad was shipping out. Grandma Merle said, "no way is my respectable baby daughter getting married somewhere way out there without me,'"and hopped on the train and accompanied poor Harold and Marjie all the way to Rockingham, N.C. where they got married, and then she stayed with them! ~ Patricia Carroll Cleveland

AFTER GRADUATING FROM THE AIR CORPS SCHOOL OF PHOTOGRAPHY AT LOWRY FIELD IN DENVER, SERGEANT H Carroll moved on to infantry school, graduating in January 1942. Next H leaped into parachute training at Ft. Benning, Georgia. After completing the requisite number of jumps, he graduated as a paratrooper in March 1942.

He was assigned to the 501st Parachute Infantry Regiment of the 101st battalion. H completed Officer Candidate School and made 2nd Lieutenant, just as Bob had demanded. They could go to the officers club together now. It was June 1942 and H was twenty-two years old.

On March 25, 1942, the 82nd Airborne Division was activated. Brigadier General Omar Bradley was the commander and General Matthew Ridgway was his Assistant Commander. The U.S. Army's First Airborne Division was now established.

At the same time, 82nd Personnel also formed a second unit known as the "Screaming Eagles" of the 101st Airborne Battalion. Bob was destined to become a Screaming Eagle as seems appropriate for a former Eagle Scout. And in September 1942, he was assigned to Headquarters 1st Battalion of the 501st Parachute Infantry Regiment (PIR) 101st Battalion at Camp McCall, North Carolina.

In October 1942, the 82nd was dispatched to Fort Bragg, North Carolina "to pursue its new airborne training… At Fort Bragg the 101st was joined by the 502nd PIR. Rivalry between the division's parachute and glider elements developed rapidly. The paratroopers were considered to be elite troops and received extra money or "parachute pay" for their hazardous missions. The glider troops, however, had duties just as dangerous but were authorized no extra pay. This situation continued through 1944, with unit commanders doing their best to keep the peace within their ranks."[165]

Bob was one of the commanders trying to keep the peace. He drew upon his vast experience as an older brother and the wise words of his great-grandmother Abby. He'd been promoted from Captain to Major on October 30, 1942. He'd received an envelope marked "RESTRICTED," and inside were Special Orders signed by G.C. Marshall, Chief of Staff.

After intensive training, the "501st Parachute Infantry Regiment was activated in Toccoa, GA on 15 November 1942 under the command of Colonel Howard R. Johnson. All members of the regiment were parachute volunteers, but only a minor fraction were truely qualified jumpers at Camp Toccoa. So, when that very arduous training was over in March 1943, the unit moved to Ft. Benning, GA to jump train all members not previously qualified. Once jump training was over, the regiment was assigned to the Airborne Command at Camp McCall, NC."[166]

1942 ended with Bob being promoted to commissioned officer on battalion duty assigned to the 501st PIR stationed at Headquarters, Camp McCall. He would eventually command the 1st Battalion of 501st PIR of the 101st Airborne Division.

The upper echelon that Bob joined had realistic expectations of the outcome of their efforts. One half of the paratroopers and seventy percent of the gliders were not expected to return home from the invasion for which they were preparing.

Back home in Sheridan, war-time food rationing had begun and people lined up to see the movie, Casablanca.

H returned to Ft. Benning with the group of men he'd been training, and awaited assignment to Camp McCall, and a badly needed break. He was going home.

In May 1943, he flew partway and then took the train the rest of the way to Wyoming. H went on leave knowing his R & R would be interrupted by required recruitment and public speaking stints.

H left Ft. Benning like the other paratroopers. "Wearing their wings, their boots polished, the trousers bloused into the boots, off they went. When they got home, they were objects of wonder to their parents and friends, obviously because of their physical fitness, but even more because of the self-confidence they had acquired in the past half year. They had been through a training course that three out of five volunteers could not complete…they had jumped out of an airplane in flight. They were elite."[167]

Women showed interest in a man wearing a paratrooper uniform. And H was a good-looking man. Compact and muscular at only 5'8, he clearly had not done as much stretching as his older brother.

Marjorie Enid Harris Carroll

He carried his 160 lbs. of muscle with confidence and a bit of a Western swagger. He was often mistaken for Bob with his blond hair, square jaw, sideways smile, and laughing eyes. That spring in Wyoming, Sergeant Carroll met and fell in love with Marjorie Enid Harris. H wrote about Marjorie with admiration:

> Marjorie was a warm, loving, shy-appearing girl. Although she was the youngest of three daughters of William and Merle Harris, she did not stand in the shadows of anyone. She deeply respected her Mother, admired her and, as nearly as possible, patterned her life and behavior after Merle. She considered herself on a level with, perhaps higher, than her older sisters in all respects.
>
> Marjorie was beautiful; she had golden blonde hair and lovely soft skin. She carried herself proudly. She had a quick, winning smile that made you want to know her. Her voice was melodious and she loved to sing. She laughed easily, not a fake giggle, but a laugh of joy. Her very touch was a caress, and I was proud to be with her.
>
> Marjorie could laugh at herself, and she could laugh with you. She loved to tell the story about sitting in the U.S.O. Club with the Aussie who told her that she had the hands of an old woman. Or about the officer who laughingly informed her that her nose pointed the wrong way. Her nose had been broken when she was quite young, when a swing came back and hit her after she jumped off.
>
> Marjorie had lived a somewhat sheltered life in a home, for the most part,

of all females. Bill Harris had died soon after Marjorie was born and Merle, with the help of her sister Hazel, made a home for Merle and Bill's three daughters and for Merle's mother, Mable.

Marj was a good student and a devoted daughter. Although she didn't show much aptitude for housekeeping, probably because for many years the family had a housekeeper who cooked and picked up after the girls. However, Merle was a fantastic cook and it was only natural that Marj too became a good cook—but never a good housekeeper.

Marjorie was warm and sexually alive. To those she loved she gave freely of herself and there was comfort in her nearness. On the other hand, there was something of a snob in her for those who were beneath her station or dignity. She could display cold disdain and aloofness.

Marj's eyes could be a warm blue or they could be a stormy gray. Her complexion was creamy smooth and unblemished but she had what she chose to call ichthyosis, a dry skin condition that she treated generously with balm barr. She used little make-up because she didn't need it. Her brows and lashes appeared almost black in contrast with her blonde hair. Mostly she wore her hair long in a page-boy, but would occasionally braid it tom-boy fashion as her grandmother had done for her for years.

In her way she was naïve and innocent yet conducted herself quite sophisticatedly and confidently.

Marjorie visited her college friend, Fritzy Walters, in Sheridan during May and June 1943. It was a typical college girls vacation and Marj and Fritzy spent their time horse-back riding, sleeping-in mornings and dancing at the Crescent till late. H's writing continues:

> The Crescent had been one of my favorite haunts and while home on leave after a very rigorous and tiring field exercise in Georgia, I went to late evening sorties at the bar. One evening, while stretching my usual Pinch and water over some time and quietly visiting with my old friend, Mike Noonan, the bar tender, I was spotted at the bar by Jerry Gwinn who was sitting with his wife, Jean Stout, and another couple, probably Bob Neighbors. Jerry rushed out and we exchanged greetings. Then Jerry, upon learning I was still single and available, then a paratroop Officer, insisted that I go into the lounge and rescue this Beautiful blonde. In spite of my protests that I was home for a much needed rest, was not interested in girls, I didn't have time, I went to the lounge to see the beautiful blonde.
>
> I wasn't too sure which of the three girls sitting alone at the table in the dark lounge I was supposed to rescue. Marjorie was dancing with Bob Burns [remember Bob Burns and the gas station incident?] when I asked if I might join them. I did ask one of the girls to dance, the wrong one, and not being much of a dancer proceeded to sit and make small talk. I later learned from Marjorie, that after she and Burns returned to the table, she had kicked me under the table and did everything she could to get me to get her away from the table. We finally wound up on the dance floor.

She was warm and cuddly and melted in my arms, an experience very new to me. With her hair on my shoulder and her cheek against mine, I was completely lost.

We agreed to meet the next evening, and the next, and the next, but I neglected to mention that I had speaking engagements—talking to various Civic Clubs about the Paratroops—and was very late in picking her up. Marjorie was pretty mad, waiting all evening at the Walters while the other girls were out enjoying themselves. Also, she had broken a number of dates with the wealthy Bob Burns, who Mrs. Walters was trying to match Marjorie up with.

Two or three nights later, after dancing at the Crescent, we had started for home and had parked on High School Hill. By that time, I was completely gone. I was not familiar with the feeling but as we kissed and held each other there was little doubt but that we were very much in love. My emotions were so strong that I ached in my desire for her. Loving her, and being of the old school of morality, I would not go beyond our kisses and caresses. But I remember saying, "What are we going to do about this?" And Marjorie's reply, "I guess we'll have to get married.

A day or two later, time is a blur now, Marj left Sheridan to go to Cedar Rapids to tell her mother and family that we were to be married. I was to follow her in a day or two and meet them." Between duties a visit was planned with the 'in-laws,' Merle Harris and her sister, Hazel Hicks, whom everyone called, "Grandma & Auntie."

Arriving in Cedar Rapids, following our whirlwind romance, I was taken home and treated royally, including wonderful apple pie. However, Auntie's opinion, "we don't believe in love at first sight," was generally shared by the family. Marj and I were not deterred and preceded with plans for an early wedding. We picked out a ring, toured Cedar Rapids and became acquainted with each other, and me with the family in the short stay before I proceeded on to Camp McCall, North Carolina.

H commanded a parachute and rifle platoon and conducted basic, physical and tactical training of his men. He was also the Battalion Intelligence Officer, the officer responsible for military intelligence training of his combat section.

Back at the Camp it was business as usual with frequent jumps, field exercises and I continued to train my newly formed S-2 section, this in the 501st Parachute Regiment.[168]

A day or two after my return I summoned up the courage to tell brother Bob of my plans to marry and thereby subjected myself to an hour-long tirade. I had no business getting married; my chances of returning from combat were very slim; I owed it to myself and the Regiment to devote 100% of my time and effort to the outfit. But, since I was committed it should be done as soon as possible. I frankly had cold feet by then and wished I could graciously or otherwise 'delay' the wedding until I returned from the war.

226 | *PHOTOGRAPHER, PARATROOPER, POW*

In January 1943, Bob had been promoted to Battalion Commander, 1st Battalion 501st Parachute Infantry. And in June, just after he turned twenty-eight, Bob reached the rank of Lieutenant Colonel. Bob had dreamed big and had achieved big dreams. He also had big responsibilities that he aimed to live up to, and he aimed to ensure that his little brother lived up to his responsibilities, too. A Second Lieutenant did not argue with the Battalion Commander! Besides, Harold had never won an argument with his older brother in his life. The paratrooper and bachelor unexpectedly found himself a groom-to-be.

"The 4th of July weekend was coming up soon and with brother Bob pushing—and arranging—and by letter and phone calls to Marjorie it was agreed that she and her mother would come to Camp McCall in time for us to be married July 3, 1943."[170]

H and Marjorie posed for this photo on the stairs. No white gown, she sits in simple skirt and sweater, he in uniform. Both are somber, serious somehow, for a wedding day. Serious because it was right back to military duties for H, and Marjorie was a bride at a time of war and understood the risks her new husband faced.

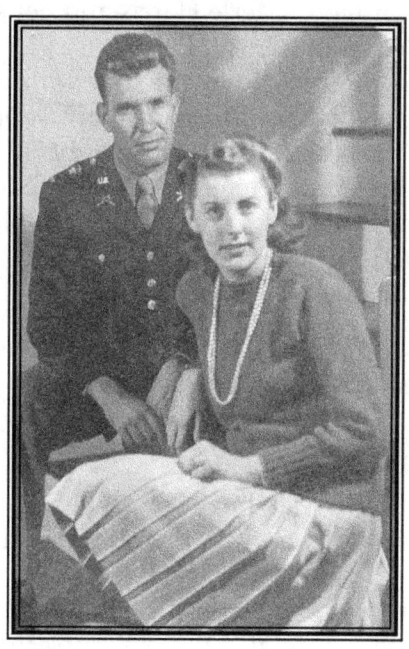

H & Marjorie wed July 3, 1943

H commanded Battalion S-2 at Camp McCall, North Carolina, from July through November of 1943. The newly married Lieutenant, in addition to other duties, continued taking photographs, including official ones:

> **12 August 1943**
>
> **Subject: Aerial Photographs.**
>
> To: Commanding General, Airborne Command, Camp Mackall, N.C.
>
> Thru: Commanding Officer, 501st Parachute, Infantry, Camp Mackall. And S-2, 501st Parachute Infantry, Camp Mackall, N.C.
>
> Request photographer and photographic equipment (4x5 speed graphic or substitute) for flight mission.
>
> Aircraft (C-47 or C-53) furnished this headquarters for Reconnaissance purposes.
>
> Mission: Photographic, series oblique and vertical photographs of field to be used during training period. Photo supplement to Map Sanford No. 31, N.C. (1941)
>
> Harold J. Carroll, 2nd Lt.,
>
> 501st Parachute Infantry S-2171

H fulfilled his duties and took a large aerial photo of Camp McCall in October or November, 1943. A series of proofs show H garbed in paratroop gear and ready to jump. He stands in the snow holding a huge camera. He is clearly going to strap the camera onto his body and jump from a plane carrying this oversized equipment! He'll take pictures as he falls. At other times, he takes pictures from the relative safety of the plane's interior. H wears an emblem depicting a replica of the giant camera with wings.[172]

His bride, Marjorie, fulfilled her duties as well. As was Sheridan custom, the hometown women's association had gathered to congratulate her.

18 June 1943

Hey Honey:

Are you ready! Well, stand up and hook up—hit the silk. I've my pipe going nicely and with a couple beers feel as though I can settle down and do this some justice —but, who knows? Meandered into the club a bit ago (it's too rowdy, right across the street from my quarters) for a beer to set off the evening—but ran into this one and that one so the beer stretched into a number, but at that have put this off only 45 minutes.

First of all let me say, this—Darling, I am the photographer of our branch of the Carrolls so just send the pictures and I shall decide (and I have long ago definitely decided) whether or not my wife to be is beautiful.

Wish I could write a love letter—full of the richest of sentiments and of moments untold but Darling, I was never instructed along that line so the best I can do is: With all my heart I alone and you and the world just stands still 'til you are where you should be—in my arms—I can't blossom and write flowery phrases nor can I quote from Shelby nor O'mar Kiegard (Is that how you spell it?) for O'mar has but one ditty which has struck a common note with me and it went something like this:

> Then to the flowing (bowl, bower, flower?)
> Did I adjourn
> My lips the sweet well of life to honor
> And lip to lip it murmured
> While you live — drink,
> For once dead you ne'er shall return

Sorry I messed it up—it does express a fine sentiment. Then too, I realize that due to my awkwardness with a pen you probably can't read half of what I write.

Do you know, we've about 20 days left, and they might well be 20 years, they surely seem that long.

Incidentally, my chum Osborn—of my old company informed of the quarters we shall likely have—with he and his wife we will have the upstairs of a house— perhaps actually, you and I will have a good, large sized room and joint bath— there will be a kitchenette of some description on the floor which we will have access to, I believe.

Honey, I adore you, please come soon, I wish you'd have been here all along.

I've nearly exhausted all the words so might just as well call it a night and try again tomorrow.

Write!
All of my Love,
Carroll

The Sheridan Star reported on November 24, 1943: "Mrs. Harold Carroll Honored." However, Marjorie may not have felt so honored. Many attended only to find out more about the outsider that the handsome, now unavailable H, had married.

Marjorie found herself hitched to a cowboy/photographer/paratrooper from a small town in northern Wyoming—a tight-knit community that didn't cater much to outsiders. The other Sheridan wives' welcome may not have been as bright as the pretty girl from Iowa could have hoped.

H Carroll

H Carroll, Paratrooper/Camera Jumper. Ref. # 173

Photo Proof. Ref # 172

Emblem of flying photographer

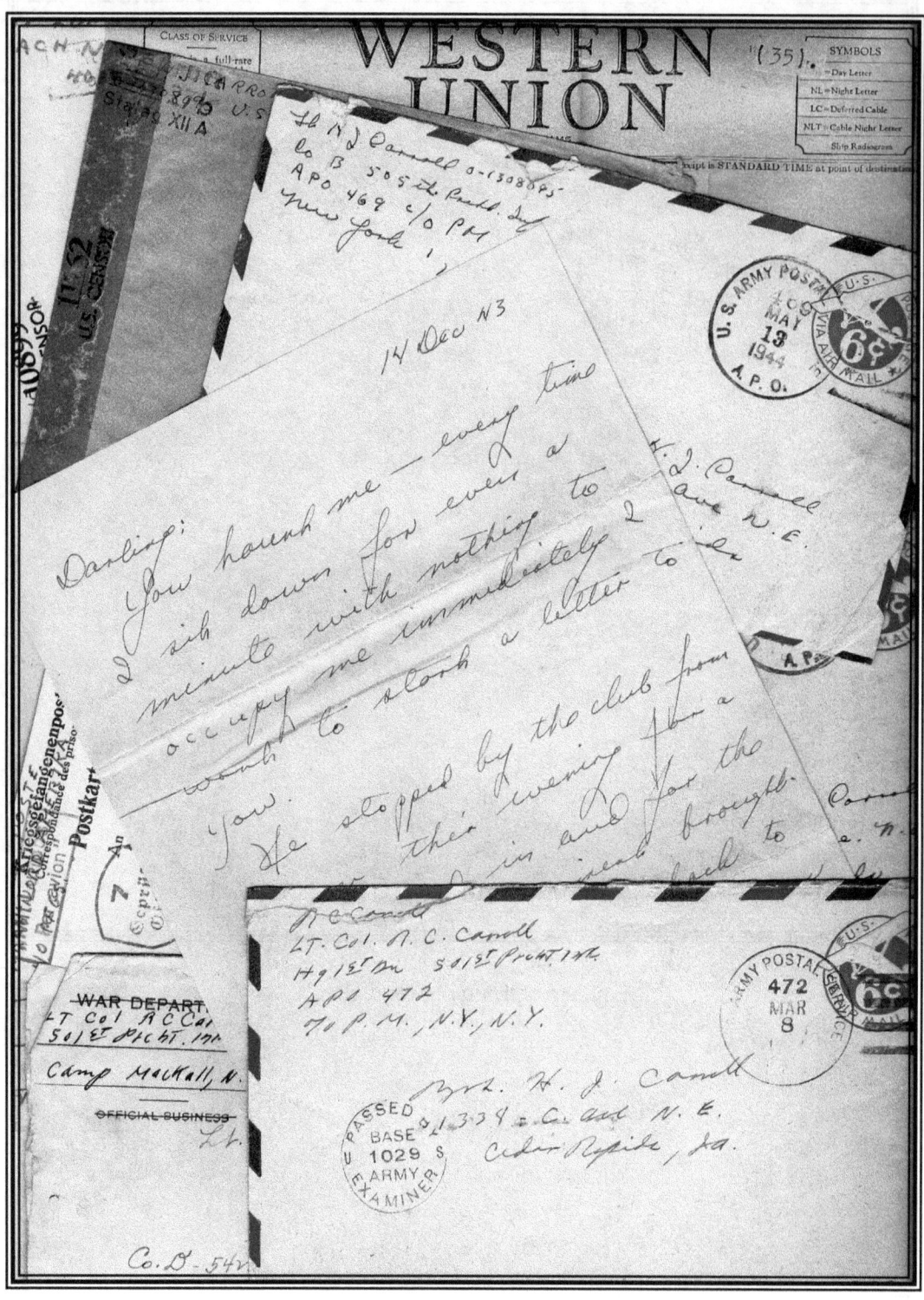

Darling: *14 December 1943*

You haunt me—every time I sit down for even a minute with nothing to occupy me immediately I want to start a letter to you.

We stopped by the club from dinner this evening for a brew and in and for the sake of experience brought a few with us back to the shack—the idea was to see how many ways we could devise to open a bottle—so far we find that you may use a GI Bunk, a foot-locker, a carbine, a door-latch, one each Sibley stove and others.

In the cockeyed mood I'm in I could Jolly well write a good letter this time if it weren't for these two other Irishmen in here.

Have to do a nasty job in the morning—processing—remember what happened to my battalion in the 501 when we were here? It happened again. Darling, you've got to get here soon—I've put in for it again—probably nothing will ever come of it but I always feel that I can't send those boys out alone.

We've been talking about army expressions—gosh, you'd enjoy it—We brought it up so I mentioned a bit of my past. Such as the origin of "Close your eyes before you bleed to death" some lad sprung that on me the X-mas morning I spent at Lowery Field a couple of years ago.

Sweet—forgive—I'm going to give it up for the night—I'll finish tomorrow…

Changed mind—after an hour of putter one of the Irish is reading a who-done-it while the other is quaking the timbers with his snoring.

You know to be very frank—I love you to beat the dickens and miss you awfully. Honestly, Darling, I feel like a homeless waif without you.

I've one accomplishment here I can always fall back on—I can say—"Well, in maneuvers, it happened thusly!" It helps.

These are two very swell Irish men with me—both as Brash as I—you'd like them.

Hey, Sweetness, I find I'm just rambling again so I think I'll close this and see about some sleep. You won't mind will you? Perhaps you will—you never wanted me to go to sleep—even when I could hardly keep my eyes open. Didn't appreciate it then but, Honey, I do now. I'd stay awake all the time if only you were here.

    None-the-less to sleep for I.
    I adore you, Write

        All of my love
        Harold

29 December 1943

Dear Marje,

Thanks very much for your swell gift. As you know a billfold is one thing that I needed and it is just the thing. The candy was very good but as is the way with candy it didn't last worth a damn. Certainly hope that you all had a nice Christmas.

I hear tell that it has been a little chilly up there. It has been so damn cold here that I have to blast myself out of bed every morning or I never would get up. After the war I'm going to spend the rest of my life sitting on top of a stove.

As for the word "soiree", Webster and I see eye to eye on most words and this is no exception. Now as for the interpretation of the thing, that is something else. As you know, there are night parties and night parties and this business about music and chow is beside the point. When I know you better I'll go into this thing a little deeper but the night parties that I'm thinking about really fit most of the situations that I call soirees except that you damn sure don't get a kiss at the end of them.

Glad that you got the flu squared away. You should take it easy for awhile until you build up your resistance from that operation. Take a few exercises or something.

Got a little pickled the other night and almost proposed to some babe but a guy shoved a drink in my hand and saved the day. Guess I need a guardian.

Remember me to everyone.
Love, Bob

FALL 1943

Surely the stories were not true; they'd been created of fairy dust or the imaginings of drunken soldiers. H was a restless twenty-three year old, full of piss and vinegar, happy to burn off his energy in a good fight. Stress was high for young men and women in service, especially high for those in voluntary service, especially high for those with a brother stationed at command headquarters, and especially high for those newly married. Was someone learning to fly?

H surely did not "borrow" an airplane with another young officer—one does not "borrow" an airplane. Certainly, shenanigans of this nature begin with something far more momentous than a simple disagreement between gentlemen, some argument about being able to see a particular thing from the air or not. Was

Written on photo:
"Two days after D-day WF still hadn't found Company B",
photographer: Harold Carroll

it visible or imagined? Who saw what? Who was right? Neither of them knew who first suggested that the only way to prove just who was right and who was wrong was to go back up in the air and see—right now.

There was no way permission would be granted for this non-military assignment to two disagreeable drunks, and so permission was not requested, but additional rations of liquor were. Then the fated take off and… of course, there was no official plane crash. If some such silly accident did actually occur, H managed to get the records of the incident squelched; perhaps with a little help from his big brother, Bob, who told him, "Go home to mother!"

Daughter Pat remembers: "Some of the stories about Dad's training in the military are vague and mysterious. Years after, I was confused whether Dad really did steal a plane while in training, and get busted, or whether that was an episode in No Time for Sergeants…? Anyway, Dad ended up in the 82nd in England, gathering for D-Day."

While H was trying to extricate himself from a sticky wicket, Hitler had found himself between a rock and a hard place. Hitler's army fared as poorly as Napoleon's army had when trying to cross the steppes of Russia during a harsh winter. He was defeated at Stalingrad by the Soviet warriors, who were used to such conditions. Hitler was finally on the defensive—a tide was turning.

"By 1943, it became clear to many senior German officers that to continue fighting a war on two fronts was bound to end in failure. It was proposed that Germany should negotiate a peace with Britain and the United States, which would then allow them to concentrate their efforts on defeating the Soviet Union. Hitler rejected this idea. He knew that the allies would insist on his removal before agreeing to a deal with Germany. Some senior officers decided together that the only solution was to assassinate Hitler. In 1943 seven assassination attempts were planned but none of them was successfully carried out."[174]

Oh, how history can turn on the outcome of a single event. Imagine if Hitler had been assassinated in 1943! How many lives would have been saved?

5 March 1944

Mrs. H. J. Carroll

Passed by U.S. Base 1029 Army Examiner.

Dear Marge,

The letters over here are few and far between, so I'm depending on you to keep up "our dear soldier boys" morale, mainly me or when I get back I'll proclaim you no sister-in-law of mine and shout stuff about never darkening my door again.

England isn't such a bad place and we are damned comfortable. The Regiment is kind of split up. My battalion is in a little town mansion which was vacated during the blitz and it has a courtyard, stables, fireplaces in every room, Ivy growing on the walls and all the rest of it, so I feel like a big slob every time I walk in. I'm billeted with a damn nice family and have a bedroom and sitting room complete with tea. I was in London last weekend and saw all the sights so now all I have to look forward to is the mission.

I haven't heard from the folks since about the first of February. I sure hope Dad is better, he has really had a time.

Haven't heard from Harold yet. I suppose he'll be dropping in one of these days.

Easy on this Uncle business, easy. Write soon.

    Love, Bob

Gen. George Marshall, Winston Churchill, Gen. Mark Clark; Lord Mountbatten, back right

CHAPTER 25

PHOTOGRAPHS, FAUX PAS, AND FOND REGARDS

Next time I'll take a parachute.
~ *A biting comment from General Lee after being injured in a Glider.*

THE PHOTOS SHOW CHURCHILL WEARING HIS CUSTOMARY WHITE-BRIMMED HAT WITH BLACK BAND. HE SPORTS A small bow-tie and a suit coat with bulging buttons, and a cane in one hand for support. He looks firm, resolute, with a serious expression on his face.

It was an early spring afternoon in the sunny south, March 22, 1943 at Ft. Bragg, North Carolina. The photos show the top brass reviewing the 503rd Paratroop Infantry including Winston Churchill, General George Marshall, Lord Montbatten, General Mark Clark, along with Bob and others, courtesy of H.J. Carroll, photographer.[175]

Bob's division continued training, and by the spring of 1943, they were ready to face their first test in local maneuvers. After the test, the 101st left to take part in a larger scale operation, the Tennessee maneuvers. "The Screaming Eagles' performance throughout the maneuvers was impressive as they demonstrated the capabilities of U.S. airborne forces."[176] During these maneuvers, however, General Lee was injured in a glider. He later remarked, "Next time I'll take a parachute."[177] This was a biting cut to the Glider pilots who were in constant battle for recognition (and pay) with the Paratroopers.

Bob had been promoted to the rank of Lieutenant Colonel in June 1943, and his father, Arthur, retired from the railroad at about the same time. Both events were photographed for remembrance and commemorated in the newspaper back home in Sheridan. High times for the railroad were coming to an end. Journalist Gertrude Spomer wrote:

> There will be no more passenger train service through the Sheridan division of the Burlington railroad.
>
> The last train, No.41 pulled in at 3.30 p.m. Monday en route to the end of its run at Billings, Mont. The cutback in service extends from Billings, to Omaha, Neb.
>
> The discontinuance came after Supreme Court Justice Byron R. White denied an application to keep two Chicago, Burlington and Quincy passenger trains running between Billings and Omaha. The application sought to stay a lower court order upholding an Interstate Commerce Commission ruling that the trains could be discontinued. The Crow Indian tribe and others opposing the discontinuance obtained a 10-day temporary stay last week.
>
> Tuesday marked the first day in 77 years in which Sheridan and other cities along the Burlington route have been totally without passenger train service. The railroad came through Sheridan in 1892."

Arthur Carroll

Arthur celebrated his son's military promotion with open enthusiasm, and his own retirement half-heartedly, although he smiles broadly for the camera in front of the Burlington Route Engine Car, his face shadowed by the customary striped cap brim. Arthur had suffered ill health for the last couple of years and now, stuck at home puttering, he had even more time to think about his aches and pains and the absence of his sons.

H had married Marjorie in 1943, and Arthur now had a daughter-in-law and grandchildren to hope for; and Bob and Art had a new sister-in-law and more ladies with whom to exchange letters.

Marjorie and her sisters, Lois and Lucy, wrote letters to other servicemen and corresponded with Bob, Art, and

H as well. It was considered patriotic to write soldiers to keep their morale up while serving away from home. Many women wrote to men they didn't even know.

Bob received a lot of letters. Some of the letters were written by two pretty, smiling women we've called "Rachel" and "Ruth", who are pictured in a photograph standing next to him in the mountains. He carried the picture of the three of them in his belongings and a framed one of Rachel he displayed on his dresser.

Rachel later sent baby pictures with her letters. Ruth conveyed to Bob how everyone was doing and kept him generally informed about his hometown of Sheridan, Wyoming. Though his reply might be brief, the letters were greatly appreciated by a soldier far from home.

When he left for England, Bob like other paratroopers was not allowed to wear his beloved jump boots because enemy spies might see them and know that an airborne division was shipping out. "They had to take the patch of the 101st, the Screaming Eagle, off their shoulders."[178]

But once they arrived, any American Paratrooper was "easily recognized by his unique calf length jump boots and sporting the hard won silver parachute badge on his breast."[179]

Bob kept very busy in England. He was involved in high-level planning and discussions with the top brass. His correspondence with his sister-in-law and others slowed as his work load picked up.

Mildred continued to write and send letters to Art. She decorated the plain, dime-store stationary with her inspired artwork, and reminders of her love in between work assignments for Joseph Platt in New York.

Art navigated the U.S.S. Sturgeon towards Japanese waters. "She sighted a seven-ship convoy with four escorts on 11 January 1944. Finding an overlapping target, she fired four torpedoes, and the cargo ship Erie Maru went to the bottom. The submarine was forced to go deep to avoid a depth charge attack and was unable to regain contact with the convoy. Five days later, she attacked a freighter and a destroyer and heard four timed hits on the targets, but the Japanese did not record the attack. Sturgeon was pinned down all afternoon by counterattacks and cleared the area at 1855. Two attacks were made on a four-ship convoy on 24 January. One hit was registered on a maru from the first attack while the spread fired at the other merchantman sent the Chosen Maru to the bottom. Two days later, she made a fruitless attack on two freighters, and the submarine returned to Pearl Harbor."[180]

Art's parents, Myrtle and Arthur, received that long awaited phone call from him after he finally reached dry land. It was then that he discovered his father was very sick.

He telephoned Mildred as soon as he could. She was in New York, far, far away, and yet he cherished the sound of her voice. Her voice was even better than her letters. And letters had kept him going during the long absences. Throughout this time of war and uncertainty, it was letters that kept strong the link between the longtime loves, those newly betrothed, and even those who'd never met.

The correspondence that the soldiers waited for, lined up to get, cherished when received, and yearned for when missing, often carried with it the images of their desire—pictures of people and places far away.

The buildup to WWII was displayed in still photos—glossy black and whites—as well as moving pictures. Pictures of men in uniform waving goodbye, pictures of the final kiss, and pictures of the gal back home—soft photographs depicting her staring off into space with a slight, mischievous smile.

Photographs of some of the top brass were taken in England in 1944. It was an auspicious occasion, "on March 23, the 2d and 3d Battalions of the 506th made a combined jump, by far the largest of the war to date for the regiment. The occasion was an inspection visit by Prime Minister Winston Churchill, Supreme Allied Commander Dwight D. Eisenhower, U.S. First Army commander Omar Bradley, General Maxwell Taylor, commander of the 101st (General Lee had a heart attack in February and was forced to return to the States), and numerous other big shots.

"The jump was a huge success. The C-47s came roaring through the sky in a perfect V of Vs...more than 1,000 men and parachutes filling the sky in a seemingly unending deluge. The instant they hit the ground the troopers were twisting out of their chutes and heading for the assembly area on a dead run...'the Boys from Currahee' [the Easy Company, boys who trained at Camp Toccoa, GA.} had made a grand impression."[181]

Queen Elizabeth smiled upon them all. Cameras clicked and flashes flashed, and the moment was captured for future generations on Kodachrome.

Gen. George Marshall, Lord Mountbatten, Winston Churchill, perhaps Anthony Eden, General Mark Clark

MARCH 1944

Soldiers filled the streets of London—American boys with their eyes wide and jaws dropped, fervently taking pictures to send home to family and friends. They swarmed a city unprepared for their numbers "…thousands of G.I.s rejoiced that the accident of war had brought them to the Mecca which otherwise they might have seen only at the expense of a life-time's savings. They were to be found, guide-book in hand, in village church and college hall…In order to pass on to them the appropriate facts that had been forgotten or never learned, the British native brushed up his own history."[182]

Bob toured London in a clamor to see everything. He was a kid again and Longon was his candy store. He dashed across the street looking the wrong direction; a double-decker bus almost creamed him. He breathed a sigh of relief on the other side of the road and looked toward the majestic British Museum. When he'd discovered the collections in the British Museum everything else paled. He was fascinated; time stopped as he explored the Egyptian wing. Glancing at his watch, he was brought back to the present. He was about to miss his dinner date! He hurried from the museum, nodding politely to a couple of English women also exploring the ancient objects, and sprinted for the taxi queue.

London was filled with women! Bob might have been in need of a guardian, as Marjory asserted to him in a letter. He was as inexperienced as any of his subordinates when it came to foreign travel, and maybe foreign women. With British men away fighting, England was crawling with single ladies.

Scholars in search of the reason for the baby boom generation (in England) have theorized that it amounts to a cultural gender difference. The theory goes that American women of the time were responsible for the amount of sexual foreplay allowed, keeping American men in check. But in England it was the fellows who were responsible for keeping the ladies' natural responses in check. Imagine when the unrestrained U.S. soldiers encountered the equally uninhibited English ladies…

In fact, many an English woman was smitten by an American soldier. According to Britain's Homage to 28,000 American dead, published by the London Times, and dedicated to R.C. Carroll, "These Americans were somewhat different, it seemed, from anything that Englishwomen had known, and difference, no doubt, is in itself attractive. They had tales to tell, perhaps not always marked by understatement, of the romantic places from which they had come and the glorious opportunities which they had left behind and to which they would one day return… They gave the impression of belonging to an earlier age of adventure; they had a glamour that certainly appealed to many young women in Britain." [183]

Although many English men had vacated quarters in London due to the call of war, there were still so many American soldiers needing a bed that England had run out of lodgings. However, Bob was made more than comfortable and in high style. He was "billeted" with a "damn nice family" where he had a bedroom and sitting room "complete with tea," he told everyone when writing home. He and his Battalion were gathered together in a little town mansion with courtyard, stables, and "fireplaces in every room." The boy from Wyoming who'd shared a bed with his brothers stretched out in comfort.

Not all Americans soldiers were so happy with their experience of England: "…there was a popular saying that England would be a wonderful country if only they put a roof on it. One American officer, it is true, was known to have spoken highly of the climate: he was neither frozen nor scorched, he said, as he was at home."[184] The quoted officer was likely one R.C. (Bob) Carroll, who spoke of the Sheridan weather with both disgust and longing. If one could just overlook the war, he thought, visiting England could be a damn fine experience.

Bob and his fellow paratroopers broke some of the stereotypes many British held of Americans. As Bob strolled through the village streets and conversed with the military as well as civilian population, a bit of the American West was introduced to old Europe. "The British learnt that the United States was not populated by versions or variants of Sam Slick, Al Capone and the Yank at Oxford…Britain had learnt the quality of its ally; the people of Britain could never again doubt American hardihood, determination or courage."[185]

Some stereotypes remained. The British discovered that, "…no American unit is complete without a Texan to extol the wonders of his State and to rejoice in the victories of Texas troops."[186] And prejudice between Americans of black and white skin color persisted seemingly wherever Americans went.

Tent CitY. Photo from canister developed in 2006. Photographer: Harold Carroll

CHAPTER 26

TENT CITY

Right is still right, even if nobody is doing it. And wrong is still wrong, even if everybody is doing it. ~ Texas Ranger saying.

On a cold morning, not yet dawn in March, 1944, H prepared to leave for Bangor, Ireland. The military had designated H as "unassigned P.O.E." (Port of Embarkment, This means hurry up and wait; just hang out at the port until your ship is ready.) He was twenty-four years old and preparing to ship off, drop off, and parachute into the unknown. All month H had waited wondering as the clock ticked toward the unknown date of his departure. Finally, in February he was assigned to overseas shipment—Troop O, bound first for Ireland, then England, and then the shores of Normandy.

H was conscious of his own "Irishness." Raised with not just a little pride in his Irish heritage thanks to Grandpa Pete, he was even more proud of being an American and a paratrooper. H was patriotic to the good ol' U.S. of A. Although raised on Irish stories and songs extolling the lovely Emerald Isle, H had also heard the gory details of poverty and struggle.

A fellow serviceman interviewed by Deryk Wills admitted, "We landed in Northern Ireland and our first camp was just outside Newtonards which we immediately named Camp Starvation. I was told by the guys that if you get put on KP's then steal some food. Of course, being a new arrival I was soon put on that duty. When I left the mess hall that night they were missing a whole ham. The men in our Q-hut thought I was a hero."[187]

James McCarroll's descendant had returned to the homeland only to discover famine was still an issue. The unpleasant conditions in Bangor, Ireland may have helped prepare the troops for worse hardships to come, but they also left H with no desire to explore the land of his great-grandfather further. Harold had experienced the American Depression and yet he and his brothers had always had enough to eat. James McCarroll, he thought, had done them all a favor when he emigrated to America.

It was early April 1944 when H arrived in England, happy for a change in scenery and food rations! It was a beautiful spring in Leicestershire. Everyone noted how unusually fair the weather was. H was a Platoon Leader for the 505th Parachute Infantry stationed with the other officers in Tent City. Here, he and his troops began the kind of training that was later described as an "intensive dress rehearsal." It lasted until dawn of the morning the invasion got underway, June 6, 1944. The men underwent the kind of "drilling" that makes family of strangers.

"Clawson and I were in the same Company, Platoon Commanders, in the 505. Clawson had been with the 505 in Sicily and Italy. I joined the outfit in England spring of '44. All of the 82nd was taken to England, with the exception of the 508th, about the time of Anzio. There were very few of the 508th that survived Anzio and they re-joined us in England April or May '44."[188]

H was the rifle platoon officer in command of parachute and rifle platoon 32 EM. He conducted basic, physical and tactical training of the platoon combat command. H was also the Battalion Intelligence Officer, the officer in the parachute battalion responsible for military intelligence of battalion training of the combat S-2 section, in addition to his related civilian occupations.

Historian Deryk Wills recounts that, "The 82nd airborne division's famous 505th parachute infantry regiment was camped in Woodlane, Quorn, known to them as 'Tent City' between Leicester and Nottingham. Many were hardened veterans from fighting in Sicily and Italy, and as 'ready day' was fast approaching passes were given out freely to the men except when night training was scheduled. Fish and chips wrapped in newspaper supplemented mess hall food and pubs were frequented. Their favorite pubs were the White Horse Inn in the centre of the village and the Bulls Head where they drank their mild and bitter beer...At the weekends they made sorties into Leicester to have a drink in the city pubs and to dance at the Palais de Dance in Humberstone Gate.

"The Grand Hotel was exclusively for officers. On their nightly tours of the Leicester pubs, drinking as much as they could, there was usually trouble. If a guy fell down drunk then his friends would take him outside and prop him up by the side of the road. The trucks going back and forth to the camp would then stop and throw him in the back. If the MP's had to step in they would do the honors for transport. Next day you would be in front of Colonel Vandervoort who would chew your ears off, but that would be the end of it.

"Surrounding Tent City was a high stone wall....If you didn't have a pass you went over the wall, and if you came back late you came over the wall. Climbing over that six foot stone wall was the only way of missing the guard on the front gate...At curfew time we caught the truck back to camp. It made two stops, one by the wall and the other at the gate.

"The first hour of the training schedule each day was thirty minutes of calisthenics followed by a thirty minute run. While no one took pleasure in this activity at 7 am in the morning, they were all in good physical condition and the only ones who had a slight problem with the exercise were the men who had been on pass the night before and had a little too much to drink."[189]

They managed to muster up in spite of the alcohol. Several training jumps took place. On Saturday, May 6th, there were reports of hundreds of parachutes floating surreally in the sky above Leicester.

APRIL 1944

It was an awkward position, not simply to be in, but also to explain. His legs were folded and sideways with his knees resting impossibly near his nose. His head was bowed with his chin on his chest in what looked as uncomfortable a pose as it was. Explanation did not come easily, especially at first.

"Yes, Sir, Colonel Vandervoort, Sir!" He tried to stand to attention and slid unceremoniously and with slapstick timing back to a lying position within the tub.

The night was dark and stormy. Clouds skimming over the moon left anyone foolish enough to be wandering the grounds completely sightless until the moonlight flashed again over the tiny English village and nearby field of tents and lighted their way briefly.

Some talk, full of bravado and envious longing, led up to the incident—as well as a liberal ration of liquor.

The street was named Main Street according to a sign tacked onto a light pole blowing over the most heavily-beaten path in the compound. The soldiers did their best to make the base feel like home. Coca-Cola signs and pin-ups of barely clad ladies were tacked up everywhere, proudly displayed.

Vandervoort was the man in charge and the man with the best benefits; he had the only bathtub in the joint. Both enlisted men and officers were required to shower, and hot water was a luxury.

It was on a dare. After a few too many pinches and water, H found his way into the General's bathtub. He didn't remember if he had actually bathed and shaved (though his previously stubbled chin was smooth) but he was thankfully dressed when—suddenly awake and aware—he tried to stand to attention.

Many sincere apologies on his lips, and a contrite look on his face, H surrendered to an almost welcome tongue lashing. The verbal punishment was in full swing when, through the window just behind the ranting Vandervoort, the dark night sky parted as the moon slipped through the clouds shining light on his buddy Carl. H could see him doubled over in silent laughter, pointing in his direction. After he escaped Vandervoort's bathroom, he enjoyed a cool rain shower and some calisthenics through the English countryside chasing after Carl.

Copy of picture on wall of officer's club

H Carroll, about 1943

CHAPTER 27

BOOZE AND BRAWLS
Give my ding-busted left ear to have you near. ~ H

TENSIONS MOUNTED AS THE SOLDIERS READIED FOR THE INVASION. PETTY RIVALRIES AND SQUABBLES OVER UNEQUAL pay paled by comparison to the troubled race relations the Americans brought with them to England. Perhaps it is true that we can see and fight injustice in others much more readily than we can see and fight injustice within ourselves.

From the Last News Edition of the Leicester (England) Mercury 3rd May 1944:

US army policeman killed in Leicester — A military policeman was one of the two American soldiers who received fatal injuries in Leicester on Monday (1st May). One of the victims died from a knife wound in the neck, the other is stated to have died from self-inflicted wounds.

The affair occurred round about closing time outside a public house in a main street near the centre of the city.

The American military policeman received fatal injuries when trying to create order.

The two men were dead when they were taken to the Leicester royal infirmary. Others received minor injuries.

All the investigations are being made by the American military authorities, who state that the matter was purely a military one, and had nothing to do with the civilian population.. In the circumstances no inquiry will be held by the Leicester City Coroner.

Corporal Abrams was killed on May 1st, 1944. Another 82nd Airborne trooper was reported killed that same day. Researcher Deryk Wills explains the events that led up to these deaths:

An African American transport battalion was stationed in Leicester, "On the night of the 28th of February war broke out on the streets of the city. The warriors had found the blacks in command of the city and the girls. There were dozens of fights and well over twelve servicemen suffered knife wounds. The interior of the American Red Cross club, opposite the YMCA on Granby Street, was wrecked.

"It was a rule that only a limited number of men received passes every night to leave the camp. On that night in question a number of 505 men were injured in the fighting. The regimental commanding officer at that time was Lt. Colonel Herbert Batcheller whose home town was in the southern state of North Carolina. As the men were forming up for lunch the next day he personally announced that, 'everyone is on a pass tonight!' His intentions were obvious. That night every available paratrooper came to Leicester looking for revenge.

"Not one word of this got into the local newspapers as it was a military secret that the United States 82nd airborne division was here in England.... Ridgeway ordered an inquiry and on the 21st March, Lt. Colonel Batcheller was relieved of his command and transferred to the 508th parachute infantry regiment. It was later claimed by General Ridgeway that no one had been murdered in the troubles, contrary to many rumors. ...On research it has been found that the violent deaths of two American servicemen occurred nine weeks later outside a public house, the Dixie Arms, in Humberstone Gate only a few yards from the Clock Tower."[190]

In a letter to H written years later, Deryk Wills explained that: "Herbert Batcheller... committed an 'indiscretion' with a young lady in Northern Ireland which somehow got to Ridgeway's notice. That was one black mark. The other was during the race riot with the Black unit in Leicester at the end of February '44. ...The Black unit was shipped out double quick, in fact they broke into an armory to get weapons and the MP's had to mount a road block just to the south of the city. Ridgway toured the city in his jeep to calm everybody down."[191]

The incident had created enough of a ruckus that those arriving afterward were filled in on every rumored detail. A lot was left up to speculation as there wasn't much time for investigation.

H arrived in England in March '44. The waiting paratroopers—a military force represented by almost every culture on earth—were still restless and bitter in May '44. But in spite of incidents of racial tension, this diverse group found enough common ground to maintain cohesiveness. Some paratroopers even adopted cultural practices not their own, like painting their faces in the Navaho fashion before going into a battle.

During the second week of May, the 101st held a dress rehearsal for its role in the coming Normandy invasion. There were three training exercises planned: Beaver, Tiger, and Eagle.

During the Beaver exercise, that took place along the Devonshire coast, paratroopers jumped from trucks instead of planes. During the Tiger exercise, they practiced capturing the causeway bridges that crossed the estuary. And during the Eagle exercise, all hell broke loose.

The Eagle exercise involved planes this time instead of trucks, and a misunderstanding caused most of the division to jump at the wrong coordinates. When the mission was accomplished, the exercise was officially considered a success. But a serious problem occurred, "Second Battalion Headquarters Company was with a group that ran into a German air raid over London. Flak was coming up; the formation broke up; the pilots could not locate the DZ. Eight of the nine planes carrying Company H of the 502d dropped their men on the village of Ramsbury, nine miles from the DZ. Twenty-eight planes returned to their airfields with the paratroopers still aboard. Others jumped willy-nilly, leading to many accidents. Nearly 500 men suffered broken bones, sprains, or other injuries."[192]

Missing the coordinates did not portend well for the future. H was shook up by the events of the day. He mentioned the incident disparagingly in cryptic language in a letter home to Marjorie. H's letter writing was beginning to ramp up, all of the soldiers were writing more as the date for the invasion grew near. Everyone knew that the letter they were writing home now could be their last.

12 May 44
Darling:
Pulled into camp this morning after a really rugged night of no sleep, too late for breakfast - too early for dinner; tired, grouchy and shocked - you don't know it but you came very near being a widow this date.
Hey Sweetness:
U.S. ARMY POSTAL SERVICE
MAY 13 1944
A.P.O.
VIA AIR MAIL
6¢
Mrs. H. J. Carroll
1338 C. Ave N.E
Cedar Rapids
Iowa
H J Carroll
MRS H J CARROLL=
1338 C AVE
(CEDAR RAPIDS IOWA)=
RECEIVED MESSAGE. K UP HOPE I FEEL HAROLDS ALRIGHT. LOVE=
ART.
26419

12 May 1944

Darling:

Pulled into camp this morning after a really rugged night of no sleep. Too late for breakfast—too early for dinner; Tired, groused off and shocked—you don't know it but you came very near to being a widow—remember this date.

Wonder if you know how very much you mean to me—some thing you can't probably say in writing, Darling, but I know you must feel it. God, wish we'd wind this soiree up so I could come back to you—It's been too damned long now and Honey I really need you.

How are you feeling, Poockie. How's Spook[193] treating you? Have you found any place to stay yet? Be thinking of some nice town you'd like to spend about a month in when I get home—where we can be alone, you and Spook and I, and can relax and start to live again. I'm going to need about a month to get back on my feet after I hit U.S. again.

Damn it all, Honey, sure do miss you and don't think Joe doesn't! Don't even worry about Joe, Darling—never. He's for you and you alone and I say this with all my heart.

Wish you'd take up the photograph bug, Honey, and send me pictures of you every so often—it helps a lot. I'd like some now and during the next few months so that I shan't feel entirely out of the picture where Spook is concerned—Would you mind?

Well Honey, I know I haven't written much but wish you'd sort of file this away for my reference when I get back.

I'll do better tomorrow when I write again.

Write -

Lots of Love, Harold

13 May 1944

Hey Sweetness:

Know what—sure love ya Babe, can I come home now? Getting doggoned tired looking at these stupid Limies—wanta get back to my bright-eyed beautiful wife. Hey Wife—how's my son?

Have no idea how this will wind up but believe you me I'll make a better stab at it than on the last one.

Had a letter from Art today at long last—answered him as an 'old married man.' His letter was very interesting. He had rec'd. the letter from you and one from Lucy[194] and was mighty happy about it you should write him again.

On duty tonight, damn it all, just about as soon go to bed. Boy—would it be nice to crawl in between sheets tonight — with you there. Sure miss you Darling.

Rec'd. your package today—you're wonderful —the candy is swell and those blades certainly are a life-saver — been hacking my face up with these lousy blades we get here; think they'll last me now—hope the next pack I get will be back in the good old U.S. and you're by my side.

How are you feeling, Honey, can't have anything to happen to you, you know.

Did you say I had two pairs of glasses? I've rec'd two cases but only the one pair of glasses.

Feel about like slugging someone in the teeth—just to be slugging them —been a couple of months now since I've had a good brawl.

Remember an old song that comes to my mind every time I think of you—which is practically all the time—'Lovely to look at, Delightful to know—and Heaven to kiss; A combination like this.' You know, bright-eyes, you'll have a hard time getting rid of me even for a minute when I once get back to you.

Honey, I've something confession—to wit, got slightly cockeyed the other night, really had excellent reason to having had a very nasty shock, just got jolly well plastered, had fun—raised hell but didn't get obnoxious. Don't really know how I did get in—or when but it was nice to blow off some steam and relax — but what a head that next morning. Came in that night with a bottle of French Vermouth—don't know where or when I got it but been sort of nursing it this evening—think you'd like it—it's something like Chablais or Souterne.

Have I groused about the blackout here yet? Sure getting awful tired of it—sure love you—stumbling around in the doggon dark having to smell your way into a 'pub'—guide your way home by a star—give my ding-busted left ear to have you near.

Are my letters reaching you all right now? Any complaints? Anything you'd like to know?

Oh yes, I'm getting so darned heavy I just sort of waddle around any more—even heavier than I was after maneuvers—and it's not all muscle now. Doubt if you'd like it.

Your way lovely you know Darling. I'm a mighty lucky guy to have you—Frankly, as my chums say seeing your picture—Don't know why I should deserve anyone like you. My old buddy Parry Schriner and I used to back off and look at each other and just wonder who in hell would ever want us. He'd say. Uglier than I—let's get a drink; and probably wind up in Denver or Cheyenne.

Seem to remember a Wed. night in Sheridan about a year ago when you were sitting next to me at the Crescent—guess it was then when I knew that you should be calling yourself Carroll. Damn it Darling, there's only one thing I want—That's you—England is beautiful, nice country, Ireland is OK—but I just want to have you close by.

There are a few expressions we have here that would amuse you. Every time these

British kids see you they rush up and say: any gum, chum? It sounds like "goom choom"—we get so we beat them to it now.

My gosh, this is becoming something of a book and still I know there's some thing I've forgotten to say. Did I remember to say I adore you?

Well, Darling, give my love to Mom, Hazel, Lucy et al—think of me after—take good care of yourself and Spook; think of me now and then.

All of my love, Harold

Yes, I'm homesick—and lonesome for you Darling.

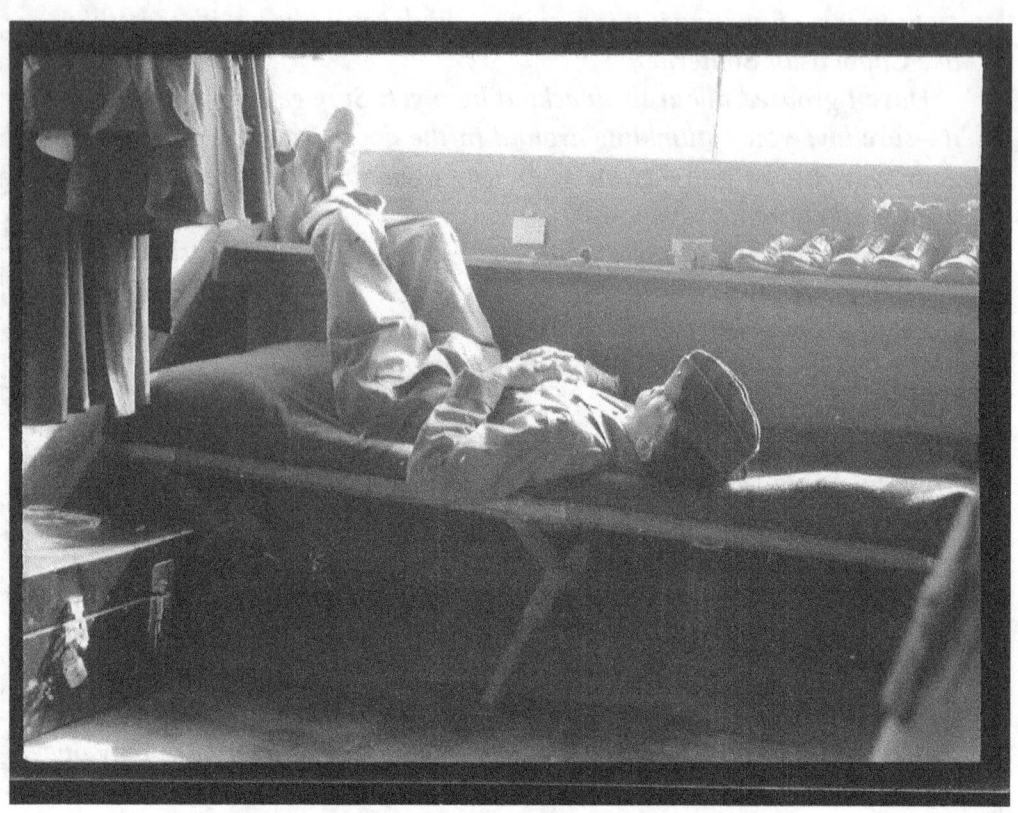

H at Ease. Photo from canister developed in 2006.

CHAPTER 28

PIG AND WHISTLE

The American soldier, in spite of wisecracking, and sometimes cynical speech, is an intelligent human being who demands and deserves basic understanding of the reasons why his country took up arms and of the conflicting consequences of victory or defeat. ~ Eisenhower.

THE SOLDIERS CONTINUED TO PREPARE FOR THE INVASION OF NORMANDY. THEY ALSO FOUND NEEDED OUTLETS TO relieve their tension, and not just through booze and brawls. This was the era of the musical, and most soldiers didn't mind being part of the entertainment. The local ATS girls (Auxiliary Territorial Service) joined with the paratroopers to produce the musical comedy, Together We Sing. Everyone came to watch. And the 82nd divisional band gave a concert in the town hall square playing to a huge crowd.[195]

Less formal entertainment was offered as well. "Robert Veria of Headquarters Company remembers the White Swan Hotel in Leicester's Market Place. The bar was on the second floor above some shops. One night several 505 men were drinking there when one shouted out, 'Standup and hook up!' Then standing on the seat by the open window gave the order, 'Let's go!' With that he jumped to the pavement, about fifteen feet below, followed by the rest of the party."

Derek Willis wrote in Put on Your Boots: "The barmaid took some time to get over the shock of seeing her customers disappear out the window so high up. From then on it became a regular feature of the evening's entertainment. On Sunday nights the Salvation Army band use to play in the market square nearby. The drummer was always ready to give a special roll on the drums as the Paratroopers did their graceful exit and fell to the ground."

Carl Clawson and H both appreciated being included by the locals in Quorn. Carl told Deryk about his experiences in The Bulls Head Inn and "the 'regulars' who were kind enough to permit our joining in their dart and cribbage games…The war time allocation of ale was in effect—but the supply was shared without hesitation." Most of the fellows called it "Horse Piss."[196] The men of the 505 consumed a lot of "horse piss." This warmed them up for the entertainment.

"We enjoyed the social life of the pub," One veteran told Deryk. "Singsongs with everybody joining in. There were no English men around; they were all in the services. The pubs were full of girls, the ATS (Auxiliary Territorial Service, later to become the Women's Royal Army Corp) from the camp at Woodhouse Eves, Land Army girls and WAAFS from the airfields. It was just like being in heaven."[197]

H had been raised with "old school" morals. He was married, in love with his wife, and quite capable of putting the brakes on socializing with women. He'd managed the flirtations of girls, young and old alike, pretty successfully. He was also admonished routinely by big brother Bob to stay focused; they could be called up at any moment. This wasn't the time to get caught up in anything, especially anything in silk stockings.

Marjory Carroll

Marjorie may have had some opportunities to stray from her marriage as well. After all, she'd only know H a few weeks before they were hitched. Then almost immediately he was gone except for short visits and letters. Marjorie, a sheltered young woman from Iowa in her early twenties, was suddenly exposed to the world. She was young and beautiful and perhaps she'd had a bit too much to drink at some lively occasion, as alluded to in Harold's following letter:

17 May 1944

My Darling:

The latest letter knocked the air out of me—it's flabbergasted I was—the three in one I mean and damned if I'm certain what type answers might be expected but now I've had my scotch ration I'll charge into something.

Holy Mack—that picnic must have been a soiree! I love my wife but hells bells. You know, Darling, I'm bout ready to toss this war into some one else's lap and race home to you—sure do miss you Poockie.

Before I get to rambling, Darling, I'd like to be very serious a moment and have just one word of caution with you: Honey, don't ever lose faith — hope; just always remember that 'no news is good news' I don't like to put anything like this in my letters but, Honey, you're all the world to me and I don't want you to be hurt so I say—just don't worry—don't ever lose faith and pray.

I hardly know just where to begin—it seems by all that's right I should have some come back for friend Carol—First of all I want to know what she meant—well lets devote this next few lines—First she says "Hi, you know what" I don't know what—What? Boy, that must have been some soiree that day—I only should have been there.

Hm—no modesty—no pants I can well imagine—I can well imagine.

If you all could have but seen the club tonight—fifty damned good men stalking around and no women, course it's an unusual evening—but reckon it would be 'field day' any day some of you lovelies put in an appearance in the GTO!

The lads here haunt the streets and clubs for women—usually a bunch of A.T.S. gals about—but—British.

Hey, Sweetness, Forgive me if I ___ this shook—I'll s_____ off a bit to Lucy and try sleeping for a change.*

Keep happy—Write

Lots of Love, Harold

---

* Undecipherable.

TWAS A FINE TIME WE HAD DOWN AT GROGANS
Irish-American Bar Song

Harold James Carroll did this so well! Great accent, etc! ~ Anonymous fan

Twas a fine time we had down at Grogans
And the five of us slept not a wink
With a fiddle to stir up our brogans
And plenty of toddy to drink.

Oh, the grog it was as free as the air is
And we managed to stow it away
We whistled and sung like canaries
Aha, who was the five did you say?

Well, meself is one, Tim Corrigan is two
Mike Strike is three, The two Grogans is four
A – er – five.

We played forty-five Mike was beaten.
Corrigan caught him at cheatin'.
They tore off their coats without waitin'
And went at it like dogs on the street.

They stirred up our blood with their brawlin'
And the fight ended up in a frey.
The five of us pushin' and hollerin'
And who was the five did you you say?

Well, The two Grogans is one,
Mike Strike is two
Tim Corrigan is three, meself is four
Oii, by dammit, I know it was five.

Oh, Pat Grogan pulled out his fiddle
Oh, Pat can sing like a loon
He drew his bow across the middle
And gave out with a bit of a tune.

Himself 'round the kitchen went dancin',
Such a jig as Pat Grogan can play
The five of us singin' and prancin'
Whooo was the five of us you say?

Meself is one, the two Grogans is two,
Mike Strike is three, Pat Corrigan

Hoy God I know there was five

Twas early daylight in the mornin'
When the party at Grogans broke up
The cock in the shed crew his warnin'
And we took one last round at the cup

But the fairest of friends must be parted and
We each of us then went his way
The five of us so happy hearted—

who-who-who-who were the five? (crying)

Meself is one, Pat Corrigan is two
Mike Strike is three, the two Grogans is four
Two, three, four…I guess there was only four.

Harold Carroll, Photographer

MAY 1944

The sky was practically cloudless, for central England. And flowers bloomed defiantly amidst the chaos. H had a lot on his mind, including a particular Artie Shaw tune he just couldn't shake. He whistled it unconsciously as he hopped off the truck in town, waving "so long" to the other fellows.

H had a pocket full of film canisters ready to be developed and a handful of mail bound for the post office and the United States. He continued to whistle as he filled out the form at the chemist shop and left the canister with the girl at the counter. She smiled pertly, and he flirted back just a bit. And why not? It might be his last opportunity.

He'd managed to carry his non-regulation Brownie with him since joining the service back in '41. He'd learned how to photograph on a Brownie, his first camera, given to him years ago by Grandpa Pete.

Grandpa Pete had suffered from a bit of a foul temper since his grandsons' departure for the war. He'd always been vocal in his opinions, and age had not cured him of that. It was an election year and Pete was an active citizen, after all. He was also a staunch Republican and a regular at the White Swan Barber Shop.

His barber, John Driskill, placed a hot, moist, bleached-white towel over his stubbled jaw. This did not prevent Pete from talking. Pete put in a plug for Dewey, the Republican presidential candidate, "Well, you know who I'll be voting for. The only one worth voting for. That's right. Dewey. And what about you, Mr. Driskill?"

Someone had left a Democrat slogan button in the shop earlier that day. After Peter's third pitch for Dewey, John slipped the Vote for Roosevelt pin onto Peter's coat when Peter was occupied with lather on his chin and looking away. John Driskill had a sharpened blade close by for protection, giving him courage.

Peter paid his tab, thanked John, and left the shop with a light step. Later, when Peter saw the Democratic button on his coat he went into an uncontrollable rage. He turned back around, reached down for the largest available rock and threw it at Johnny's shop's front window. He was so angry he shouted, "You can call me an S.O.B., but never call me a Democrat." (He didn't pay for the window either!)[198]

Lieutenant Art Carroll returned home to Wyoming from the South Pacific in the spring of 1944 on emergency leave to visit his dad. Arthur had suffered a heart attack and was in the hospital. Both Bob and Harold were in England preparing for the invasion. Responsibility for their father rested with his middle son. Of course, Grandpa Pete was there to assist.

And Peter was in a foul mood. He didn't think it right or fair that his son, Arthur, should be in worse physical condition than he himself. Arthur's condition finally improved and he was sent home from the hospital within a week and Peter breathed a sigh of relief.

Lieutenant Art was ordered to report to the naval base at Newport, R.I. The letters he'd received from his darling Mildred had kept him amused and inspired during the many months navigating dangerous waters. He followed their scented trail to NY from Newport when he had time off from his heavy work load as Trouble Shooter and the only officer over one hundred and twenty-five men.

It was springtime in New York. Art had taken many trips back and forth between Newport and Manhattan, but this train ride seemed the longest. He had finally found the courage to propose—if Mildred could only find it in her heart to agree to marry him.

He felt a bit of trepidation and asked her shyly, but he was certain of her answer. Of course, she would say 'yes.'

Milly's sweet brown eyes lowered, "No," she surprised him. "I'm sorry, Art. I like my job and my life here in New York and I don't want to get married," she flatly refused.

Art returned to the base more let down than he'd ever felt. But he was also stubborn. Art wasn't about to give up. Mildred simply needed to be persuaded; he had been away too long. She wasn't getting rid of him that easily. He could be persistent.

Mildred & Art, Wedding Day

And Mildred could be determined — determined that she didn't want to marry. Not yet. Her career was just beginning. She had a fantastic life, a dream life! She had the world on a platter... but she had the "flu," a lingering "bug" she just couldn't shake. Weeks went by. Her roommates watched her walk around in a state of lethargy in her pajamas, day after day. Finally, Loraine said, "Oh, you don't have the flu at all, you're in love and won't admit it."

Art was prepared to ship out at any time on the U.S.S. Stag as navigator. The U.S.S. Stag was a water distilling ship sporting new, crucial technology. Mildred had refused him so adamantly that he was shocked by her question the following weekend over coffee. "Well, when do you want to get married?"

He backpedaled, practically choking, "We should wait until I return, Milly." He was concerned now, concerned for her, knowing he was going to ship out again soon. Knowing what it was like out there; "What if I don't come back?"

"You already talked me into it." Mildred stood her ground, her petite form unwavering even though she was the one who hadn't wanted to get married to begin with. She had a fabulous job, great friends, she loved her life. She also loved Art whose gentle six-foot frame grabbed her up in a spontaneous bear hug and swung her around the room.

Art had no time off so they were married in Newport, Rhode Island on May 27, 1944. Loraine was Milly's maid of honor and Art's best man was Officer Clark Tothrow. Mildred's dad couldn't come to the wedding so her boss, Joseph Platt, gave her away.

"The Platts became very involved in my wedding," Mildred remembered. "The wedding reception was held at their summer home in Little Compton, R.I. June Platt's garden was all planted in blue and white flowers…beautiful and perfect for a wedding." Mildred recalled, "My mother and sister, Jean, came from Oklahoma. And Harold Town, my minister from St. Thomas Chapel in New York, did the honors in Trinity Episcopal Church, one of the oldest Episcopal Churches."

Roommate Pauline's wedding gift to Mildred and Art was a painting of the East River and the Beekman Place and 57th Street Park, a favorite haunt of the pair and a place they might be seen strolling hand in hand.

Mildred & Art

B-company, a few days before D-day, photographer: Harold Carroll

CHAPTER 29

THE LAST PHOTO OP

Some of you will die, this participation in world history may be just a thought that you can take with you. You are assured by me that you will be on the winning side. ~ General Ridgway speaking to the troops.

For some, it's a rabbit's foot. For others, a letter so often read the words are faded, or a bent and worn photo of a sweetheart. H grabbed his old Brownie. There's comfort in a camera. There is, if nothing else, a sense that in the future, someone—perhaps you if you're lucky—will be looking at the developed photos, now memories.

"Those who were able to sleep were awakened early on the morning of May, 29th. They gathered their gear silently, tucking last minute souvenirs and good luck pieces into clothing, or quietly cursing the precious items lost to late night gambling sessions. The 505 paratroopers hauled their packs up in the still and misty dark and climbed aboard trucks. The paratroopers left for two airfields to prepare for the D-Day invasion."[199]

H gathered and double checked his equipment and looked in on his men. "'That morning as we marched out of the camp to meet the trucks, the street was lined with our friends. The people of Quorn had a much better 'grape-vine' than we did.' That was Harold's last sight of Quorn — he never made it back."[200]

The soldiers stationed in England, now preparing for the invasion of Nazi Germany, were on a secret mission. And so H, the intrepid photographer, took the photos on his personal Brownie rather than the military photography equipment he'd grown accustomed to over the last few years. It seemed like the perfect photo opportunity. After all, it was a time of leave-taking, a time for goodbyes and heartfelt good lucks. They all knew the odds. Many would not return. Some of the boys with innocent, smiling faces, looking sheepishly or rakishly at the camera were ghosts already.

No matter the ribbing, H took the shots in reverence. He always took photography seriously. Here's a shot with some jolly fellows from B Company standing together in Tent City. A tarp is spread out in front of them with belongings laid out in preparation for the next jump—the one that isn't practice.

They stand resolutely in another, lined up preparing to board a plane, wearing most of their gear already, each paratrooper carrying about 150 pounds in weight, including the big packs on the ground at their feet.

And here's a shot of the planes, the men standing in front of the C47 aircraft at Ashwell airfield in Rutland, smiles that reveal unease.

One photo shows a young man intent on writing a letter. He wears his hat slightly askance and leans over the table intently working on the second page, presumably to his sweetheart whose merry eyes and bright smile are framed on the shelf behind him. You can almost make out the words of the letter…

H took the film of the smiling and intrepid 82nd to the local chemist shop in the village. None of the men knew the exact date of the invasion; count-down was already at ready day plus. H dropped off the film just before June 6, 1944. Forty-six years later the pictures resurfaced.[201]

The Cottesmore and Spanhoe airfields were both used for the invasion. Cottesmore boasted seventy-two C-47 aircraft ready for flight; Spanhoe, fifteen miles away, readied forty-eight C-47s. Deryk Wills describes the last minute preparations:

> At Spanhoe, 864 Paratroopers crowded into a hangar for their sleeping accommodation. In all 2095 men of the 505 were standing by.
>
> Craps and poker games were in full swing to help pass the time. The stakes were dollars and the recently issued French invasion money. Fortunes were won and lost in the seven days they were cooped up waiting for the word 'Go'.
>
> A summer gale hit central England that night and the 505 rose to the sound of rain. The wind gusted strongly enough to postpone the jump. Tired men, taut with adrenalin, waited…
>
> Some of the 505'ers chose to blacken their faces. Orders were to have their hair cut to a half an inch in case of head wounds. Everyone was checked to make sure that the two metal identity tags held by a thin chain were around their necks. This was a long running joke amongst

the 505 as they were always losing them. Many a local girl around Quorn and Loughborough, and the ATS (British Army) girls from Woodhouse sported jump wings or a set of identity tags.

At ten o'clock as the daylight began to fade the order 'Chute up' was given. Equipment bundles had been fixed on racks under the planes to be dropped at the same time as the men. The parachutes were laid out by the Supplymen, everybody picked his main and reserve out of the pile and proceeded to adjust the harnesses.

Each man was now carrying at least 150 pounds in weight. On the command 'Load up' the Troopers made for the tailplane of the C-47 to relieve themselves—a paratroopers ritual: not the easiest thing to do with all that equipment and the parachute harness on. Willing hands helped the bulky bodies up the short ladder and through the door of the aircraft. After ten minutes in the plane there was always somebody who wanted to head for the tailplane again.

Just before dark, about 21.30 hours on 5th June 1944, the English countryside heard the sound of aircraft engines warming up. Large numbers of C-47's from numerous airfields across the country took off, circled until they got into formation, and headed across the English Channel to Normandy. The invasion of Hitler's Europe had begun.[202]

The wait was over. The time they had trained so hard for, that had filled their waking hours and created nightmares, had come. It was time to check gear. Stand up and hook up.

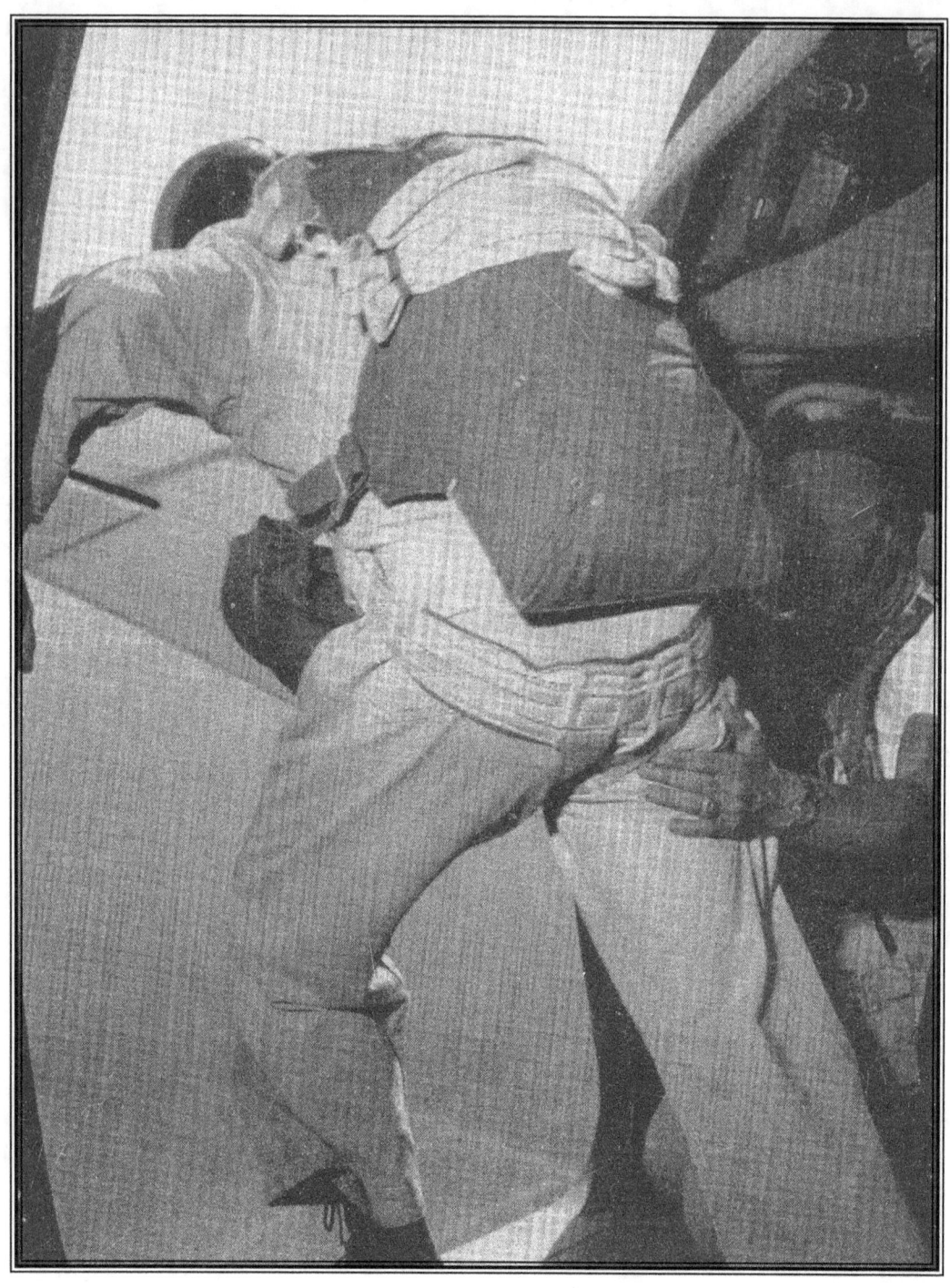

The Jump, photographer unknown, from loose photo card circa 1942

PART IV

H Carroll, Ready to Jump

CHAPTER 30

SIGNALS IN THE DARK

The code name for the 82nd Airborne drop was 'Boston.' At 11 o'clock the C47s took off. In the darkness of the Normandy countryside identification was going to be made by the password 'Flash,' to be answered by 'Thunder', and by little toy crickets which the men clicked between their fingers and thumb.
~ Deryk Wills

June 6, 1944, the Invasion began. "A 23-year-old officer commanding a platoon of parachute infantrymen that all called him 'H', Carroll was on his way to the skies above Normandy."[203]

"Going out across the channel, it was beautiful in a way." H spent much of his time standing in the door surveying the massive naval forces gathering below. "From the air it looked like you could walk across the channel from one boat to the other."[204]

Parachutes and gear in place, he and his men waited anxiously for their signal to jump.

"It was very dark and very wild, confusing, erratic. There was flak and ack-ack and flares and the sky was absolutely full with shells from Germans trying to shoot our planes down," H reported. "We had nothing to do but be sitting ducks waiting for our signal…I know that all of us were absolutely scared to death. But I had quite a number of men counting on me, so I really couldn't think of my own feelings too much." H jumped with his platoon into the black, blazing night.[205]

"The main problem in those first few minutes of the 6th June was the low cloud which obscured the target areas…Because of the flak and nervous tension the pilots did not always throttle back and reduce their air speed as they approached the Drop Zones. This meant that the troopers had to jump out at speeds in excess of 150 mph. The opening shock tore off musette bags, binoculars, watches and anything else which was not securely fastened."[206]

Pilots could barely see through the fog and flashes of anti-aircraft fire, and poor visibility forced some of the planes to break formation. Paratroopers from both the 82d and 101st Airborne Divisions missed their landing zones and the men were scattered far and wide.

The overall mission required the paratroopers to knock out bridges and secure roads to make sure that infantry troops landing by ship on Normandy's Utah and Omaha beaches could make it inland through the Nazi's barricades. If the Infantry failed, then the paratroopers would be left stranded behind enemy lines.

The mission of H and his men was to provide perimeter defense for a battalion crossing a small river. "Our mission, at least my Battalion, was to seize the river crossing over the Merderet almost due west of Ste. Mere-Eglise. We did encounter very heavy flak and Ack-Ack as we swung east from the coast. However, instead of going east from approximately Carteret our C-47s cut east on line with Bricquebec. This resulted in our being dropped very near Valognes which was a heavily concentrated area."[207]

H landed into a nest of Nazi soldiers about ten miles from the point that was originally selected, and far from most of his men. His first priority was to regroup, determine the location of his platoon, and carry out the mission, and, of course, to avoid getting killed. Enemy fire was all around—H took hits of heavy shrapnel in his neck and was wounded in the knee.

Separated from his platoon, he organized other Allied stragglers and tried to carry out his orders. "I gathered men from the British, the 101st, the 82nd airborne and Free French; in the middle of the night, dark as pitch, signaling with our crickets… I managed to get a pretty good force under my command," H recounted.

Bands of separated troops congregated and fought together against the Nazis. When the plan to capture the bridges over the Merderet failed, there were a number of groups who "fought independently—in some cases for four or five days. These isolated groups contributed in some degree to the accomplishment of the divisions' missions, though they carried on what amounted to fights for survival rather than battles for planned objectives."[208]

H and his makeshift force resisted for three days, until on June 8, 1944, the Nazis captured them and H was taken prisoner.

"After our capture, which was somewhere between Montebourg and Ecausseville, we were taken to Valognes. We were held in a large fenced-in enclosure (which was a target of our Navy guns); then marched to Cherbourg. The Germans during this first capture couldn't believe that we were part of the invasion force and interrogated us very thoroughly. These being 'Front Line' troops they did treat us with some respect—not as we later experienced with Rear Echelon troops," H recounted.[209]

Bodies covered Utah Beach, turning the white sand red with blood. The allied forces storming the beach encountered land mines and barbed wire rolls and heavy artillery fire. But U.S. and British forces possessed tenacity and obstacle-busting tanks. And they had a seemingly endless stream of boys and men who climbed over each other's fallen bodies in their rush to meet the enemy.

Col. Vandervoort broke his leg in the jump the morning of June 6, 1944, "but he ordered a medic to lace up his boot tightly so he could keep fighting as his troops turned back a German counterattack on the Allied forces landing on Normandy beaches. He used a cane to stay in combat until the beachhead was secure."[210]

Official Report — D-Day — Operation Neptune

The 101st Airborne Division first saw combat during the Normandy invasion – 6 June 1944. The division, as part of the VII Corps assault, jumped in the dark morning before H-Hour to seize positions west of Utah Beach. Given the mission of anchoring the corps' southern flank, the division was also to eliminate the German's secondary beach defenses, allowing the seaborne forces of the 4th Infantry Division, once ashore, to continue inland. The Screaming Eagles were to capture the causeway bridges that ran behind the beach between St. Martin-de-Varreville and Pouppeville. In the division's southern sector, it was to seize the la Barquette lock and destroy a highway bridge northwest of the town of Carentan and a railroad bridge further west. At the same time elements of the division were to establish two bridgeheads on the Douve River at le Port, northeast of Carentan.[211]

101st Airborn, Bob fourth from left

CHAPTER 31

SCREAMING EAGLES

These were the kind of people you would want covering your back in a barroom brawl or protecting your flank in the next foxhole. They were the kind of people who would fight like madmen, then after it was over look down at a dead friend or relative and cry like babies. ~ James Webb, Born Fighting.

BOB AND THE MEN OF THE 101ST FARED EVEN WORSE THAN H AND THE MEN IN THE 82ND. FORCED TO CHANGE direction and speed because of the heavy anti-aircraft fire, Bob must have had a sense of déjà vu as his battalion struggled to stay on course that erratic morning of June 6th.

"The serial consisted of 45 planes. Of this number, only 18 dropped their sticks on or near the drop zone. The rest were badly scattered with some planes giving the jumpers the green light several miles south of Carentan."[213] (So bad was the situation for this battalion that it had to be reorganized on D-Day plus 2, June 8.)

The Eagle was an apt symbol, with its ferocious screech, powerful talons, and acute eyesight for both the 101st Airborne and for Bob, who had achieved the Eagle rank in scouting. He had been forewarned of trouble by the Eagle preparatory exercise during the 101st dress rehearsal in which the mission had succeeded but the troops were scattered.

The paratroopers of the 101st were promised reinforcements. At dawn June 8th fifty-one gliders were scheduled to land. Many crashed, however, and several soldiers were killed, including the assistant division commander, Brig. Gen. Don F. Pratt.

101st Airborn Command Personnel, Bob Carroll back row, second from right

Bob commanded the 1st Battalion of the 501st PIR. Their mission was to capture the locks at La Barquette and to destroy the railroad bridge that lay northwest of Carentan. The Germans controlled the flooding of the surrounding countryside by use of the locks and the bridge was used by the Germans to bring in reinforcements. The Germans flooded the fields which caused some paratroopers to drown and enabled easy capture of others.

"The men of the division, however, persevered and proceeded with their assigned missions as best they could. By nightfall soldiers from the 101st had secured the beach exits in their zone and contacted the landing forces of the 4th Division. The Screaming Eagles also controlled the La Barquette lock, but could not secure crossings on the Douve River. The following day 101st elements attempted to advance in the division's southern sector, but made little progress against heavy enemy resistance near the village of St. Côme-du-Mont."[214]

The military reported Lt. Colonel Robert Collin Carroll and 2nd Lieutenant Harold James Carroll missing in action and presumed dead on June, 8, 1944.

"There was some confusion as regards my status following D-Day in that my Platoon Sergeant reported me as killed in action when the Regimental Casualty report was given after my regiment was returned to England."[215] But H was still scrambling for survival.

Raymond Paris, a local Frenchman, recounted to Deryk Wills that: "during the night of 5/6 June after the parachute jump, as I was coming back in the darkness from the house which had burned in the square, we found in the middle of the street, twenty meters from our home, a parachutist of the 82nd Airborne who appeared dead. We carried him a few meters away to the edge of the square and covered him with his parachute.

"The next morning I went back and I was amazed to discover that there was no body under the parachute. Nearby a parachutist was sitting on the steps of a truck left behind by the Germans the night before. He spoke French and told me he was from Louisiana. I told him that the night before we had carried one of his friends who had been killed.

"He burst out laughing and told me that it was him. He had been wounded in the shoulder by a bullet and that he had shammed death in order not to be killed by the Germans. When he heard us coming, without knowing if we were French or German, he kept playing dead."[216]

101st Airborn Command Personnel

8 JUNE 1944

H was dirty and hot and hungry. The hills in this part of the French countryside rolled endlessly towards the horizon, the road running through them went on seemingly forever. He'd tasted nothing but dust for miles while marching to Rennes as a Nazi prisoner of war.

The Nazi soldier had been eyeing him for some time, his hawkish face partially hidden in shadow. H was unable to get a good look at him with the glaring sun reflecting off of the soldier's long brimmed hat. But the intensity of his stare could be felt without seeing his eyes. H had done his best to ignore the unwanted attention of old hawk-nose for too long now and was getting edgy.

He had been wounded by shrapnel on that first fire-filled night, just above the left knee cap, and the wound was beginning to fester. He found himself falling back toward the rear of the formation, limping a little as he nursed his knee. His neck was sore and bleeding, too.

The Nazi had no trouble catching up to him. He looked H over carefully before announcing, "I've seen your brother. He is with the group of prisoners just ahead. They are marching to another camp." He spoke excellent English quietly in a matter-of-fact tone.

"My brother? How do you know he's my brother?" H asked, ever skeptical.

The Nazi laughed aloud. "You are twins, no?" He moved forward, leaving H a bit more confident about the well being and whereabouts of his brother, Bob, as he scratched at the shrapnel in his neck and wiped the blood on his filthy trousers.

2nd Mapping Sq.
Felts Field, Wash.

CHAPTER 32

IF AT FIRST YOU DON'T SUCCEED, ESCAPE, ESCAPE AGAIN...
We, at that time, exchanged helmets with a couple of Enlisted men —the Germans only counted the Officers. ~ H Carroll

The prisoners of war were marched under strict guard and without mercy from Valogne to Cherbourg arriving on Jun 10th. A few days later, June 14, they continued on to Bricquebec, France. H and Clawson hung together. "There were quite a number of guards all along the column."[217] Nazi soldiers prodded them with fists and sticks and slurs. Some of their insults were unintelligible, and others needed no translation.

H and Clawson were good friends back in England where they'd first met. The experience of being prisoners of war together cemented their personal alliance and they became a team, resolving to elude their captors. But H and Clawson found it hard to reach Allied forces. "It was very early in the invasion and there were no lines set up, no rear echelon, no place to go. So ultimately we were recaptured and then transported on down to Rennes, which was the first major prison camp that I ever saw."

Their march to Rennes began on June 17, 1944. H left Bricquebec and spent one night at St. Sevier Le Vicompte and another night at Le Haye Du Puits. And then began the "starvation march" to a point approximately 100 miles away from Bricquebec.

"The march was well named, for the Germans had made absolutely no preparations for food for the prisoners. They marched at night to avoid Yank capture, and during the day stayed in French homes or barns. Sometimes, they took over a French kitchen and were able to hurriedly prepare soup or some other concoction that kept them going on the five-day march."[218]

On the fifth night of the march, June 20, 1944, H and Clawson escaped again, this time near Coutances. "That night Clawson and I hid in the area between the ceiling and the peaked roof of a mansion the Germans had placed us in during the day. We, at that time, exchanged helmets with a couple of Enlisted men—the Germans only counted the Officers—and, well after dark, after the Germans had moved out the column, we left the 'Chateau' and made our way north—traveling only at night. That period of time is rather a lengthy account."[219]

By careful calculation and by traveling only at night, H and Clawson managed to escape capture for 10 days, but again, lack of food proved to be a tough proposition. They remembered only three meals during the 10 days.

It was while trying to sneak through a thick German forward wall near the battle line that they were spotted and recaptured by the Nazis. They were seized at St. Jean de Day, near Le Hayes du Puits, on June 30, 1944, trying to find their way to the swamp that separated the Allies and the Germans.

H and Clawson faced their captors with a good deal of anxiety, not certain what the policy would be on escaped prisoners.

13 JUNE 1944

D-Day resounded around the world. Newspapers marked its significance on the front page in bold print, in many languages. Myrtle received Western Union notices in misspelled, abbreviated English that her sons, Bob and Harold, were both missing in action and presumed dead.

Arthur had never recovered from the heart attack he'd suffered back in May, and he gave up the will to live. Myrtle believed that the loss of his sons killed him, although she'd never allowed radios or newspapers anywhere near his hospital room. She even hesitated to tell him about the telegram.

When her husband died, Myrtle was devastated. Her family had been taken from her practically overnight, her strong husband and her strapping boys. Arthur was only fifty-three when his heart gave up on June 13, 1944.[220] Myrtle had always had her father George G. to look up to; and her husband Arthur had been her rock, and she'd raised three strong sons. Now she was alone. The mountains closed in on her.

Myrtle was unable to reach middle-son Art, and so she called his wife Mildred in New York. Milly took a no-nonsense approach to the situation and telephoned the naval base. She requested Art to be relieved immediately for his father's funeral.

"Don't you know we're in a war, lady?" the first officer she spoke to complained rudely.

"I'd like to speak to your commanding officer," Mildred persisted undaunted. Finally, she got the Commander on the phone.

Art was potentially Myrtle's only living son and her husband had just died. The Commander was very polite and arranged for Art to fly to Sheridan to be with his mother. He also sent Mildred by train to Cheyenne, Wyoming and then by plane on to Sheridan to be with Art and his mother in their time of mourning.

Mildred found herself in the midst of Carrolls—about two hundred head—who all converged for Arthur's funeral and reception following. The families of both Peter and George Carroll were present. There were some rivalries between the Carroll clans, but despite their differences the wake was mostly peaceful. Myrtle, silent and stoic, tried to stay busy; she was so thankful that Art was home and safe and sound.

Peter Carroll spotted Mildred about the time that Mildred noticed the slight gnome of a man approaching her with a sly smile. Meeting Art's family the first time at Arthur's funeral was awkward, but Peter put Milly instantly at ease. He was not much taller than she was, but he offered her a warm welcome and announced in his thick brogue, "Ah, I couldn't have picked a better one if I'd picked you meself."

Myrtle dealt with her grief by digging dirt. Working in her flower gardens, she told Mildred, was life affirming. "The flowers bloom, I can count on that."

She reminisced about Bob, Art, and Harold to Mildred as she planted another bulb. "I fussed and complained to the boys about the dirty woodwork for years! I was always trying to get them to wash their hands. If they'd just wash their hands the woodwork wouldn't always need cleaning!" But after Bob, Art, and Harold were all gone and Arthur buried, she lamented, "Somehow the woodwork still has dirty handprints! They appear as if by magic."

And Myrtle talked about Bob, who could also be very fussy. "One day I was baking bread, and after cleaning the pan and breadboard, I put the crumbs and doughy scrubbing pad out for the birds. Well, Bob returned home and saw the scrubbing pad in the yard and was horrified. 'Mother, what on earth is this doing in the yard?' he complained. 'I'm feeding the birds while I bake bread,' I explained. He just shook his head, exasperated."

"Bob is so very neat and orderly," Myrtle reminisced, "and so structured about educating himself. He has a wonderful library, and he's very well read, you know, even more than Art. He diligently followed the recommended reading list provided by his uncle, Robert E."

Mildred listened patiently. Her mother-in-law seemed to need to talk.

Finally, Myrtle confided in Mildred, "You are practically the only person I am willing to talk to about Art at all." Myrtle did not feel like explaining her motives regarding Art's return to service to anyone. His father's funeral had barely ended before family members began pressuring Myrtle about keeping Art out of active duty. Some pressured her frequently and relentlessly.

"You have the power to prevent Art from returning to active duty," was the mantra heard from kitchen to front-porch.

Mildred watched Myrtle stand stoic, refusing to budge, but they kept after her.

"You could write a letter and request that Art not be sent overseas," her sister told her again on the phone. But Myrtle steadfastly refused. "I can't prevent Art from going. It is God's will." She was sparse with her words but firm. "He could be killed on the streets of Sheridan." Myrtle insisted, "It's not for me to decide where he should be. Art chose submarine service. That's his chosen path. How can I keep him from his chosen path?"

Art returned to Newport and duties as usual. More men and officers arrived every day making housing tight, but Mildred and Art were able to rent a small house on Bliss Road. But blissfull togetherness would not last for long. It was just a short time after the funeral when Art was ordered to report to Tampa, Florida for the commissioning of the U.S.S. Stag, a water distilling ship in need of a navigator.

"I was wearing a wool suit when I arrived in Miami!" Mildred, overly warm in the summer sun, carried with her the top of their wedding cake carefully wrapped in waxed paper and secured in a hatbox tied with string. "We went by train and there were about a hundred and twenty-five men, and me, Mildred!" When the string on the hat box broke and the cake rolled across the floor, "all the men clapped!"

"We were billeted at a hotel in Miami and Art got food poisoning." Mildred was determined to rent a more suitable place. "Bull-headed, I finally found a room with Mrs. Adamo, the wife of Frank Adamo." Frank had been in the Philippines with no medicine and many wounded troops. He put the injured soldiers out in the sun for a limited time to help cure infections and was credited for saving many lives. Mrs. Adamo's daughters were away with their husbands who were also in the service, and so she was happy to rent some rooms. "She had very strict rules about the kitchen because of the ants and critters," Milly remembered. "And she advised us to take short sun baths to get used to Florida and the heat."

"Practically as soon as the ship was commissioned Art told me to go visit my brother and parents. And there was to be no discussion. That was the end of the conversation."

"Don't ask questions, just do as I say," he insisted. "Go visit your brother Ralph in Birmingham, then go home to Oklahoma. Please."

Mildred thought this a very strange and discouraging request. No questions! Unknown to her, Art had requested to be stationed in Galveston, and if Mildred went home she would be close enough to visit when he transferred. He didn't want to say too much so as not to disappoint her if his plans failed. And so, with nothing more than sideways glances and questioning eyes, she took a sabbatical from work, and left.

"I did as you asked," Mildred said when Art telephoned at her parents' house in Ada, Oklahoma.

"Good. Now you can come to Galveston." She could feel his smile through the miles of telephone line that separated them, but now far fewer miles than there could have been, and soon they would be together again.

Mildred's parents took her to Galveston, and they rented a cottage. "It was a good family time. Art was gone for periods of time to do what Navy men do to ready a new ship." Milly had time to swim, read, and explore, and eat leisurely meals with her parents who she'd rarely seen since she became a New Yorker.

Art rented a hotel room by the week for his extended stay. The maid in his hotel pointed out one morning that Art's room held two beds, and two people left the room, but "you'se only use one bed!" John was conceived in Galveston that September.

Soon the U.S.S. Stag was ready. Art and Milly said good-bye at the dock and resolved to write every day. "I went back to New York after a couple of weeks visit.

He went on out to sea." They stayed in touch as best as they could. Sometimes there would be two or three letters at a time and on other days there were none. Mildred learned to read the newspapers carefully.

The Stag was a water distilling ship, a new thing. Nothing had turned sea water into drinking water before. "After he left, I would follow every word in the newspaper and I could almost imagine where that ship was going to go to give the men fresh water." If there was a landing of our troops Mildred guessed, usually correctly, that the U.S.S. Stag would not be far behind. She knew where Art was by reports of liberated men. Where there were lots of men, there was need for clean water. Although this was never confirmed, it certainly made her read the news more!

John was born while Art was still in the Pacific. Mildred took the new baby and went to Sheridan to stay with Myrtle and await Art's return.

A MOTHER'S PRAYER
Reuben Coldsmith

Somewhere across the infested sea,
Serving the cause of liberty,
Dear Lord, is my brave son.
I know not how he fares tonight.
But keep him ever in Thy sight,
For I have but the one!

Thou gavest two to bless my life,
But one fell early in the strife –
My first-begotten son!
He with his comrades marched away.
And then a message came one day—
Dear Lord, I have but one!

Then while so sorrowed and bereft,
Though only he to me was left.
I could not say him nay,
Who urged that duty bade him to –
His country called, he said, and so
How could I bid him stay?

But, O, dear Lord, 'tis hard to bear,
With not another child to share
The hearth when day is done!
I can but Kneel to Thee and pray;
Be with him through the hellish fray
And keep Thou safe my son!

I know the cause is good and true—
I know that suffering must ensue
Before the fight is won.
Yet bear with me, O Lord, this night;
Be Thou my solace in my plight
And guard my only son![221]

B-company, a few days before D-day; Carl Clawson second from left,
photographer: Harold Carroll

CHAPTER 33

STARVATION HILL

The Major mellowed and offered us sips from a jug of Aquavit. As we lounged there watching, two or three of our planes were shot down. The Major gloated that his pilots had shot down 25 of our planes that day. Clawson informed him that: "Shit, we made 150 of those planes [this] very day!" ~ H Carroll

H CONTINUED TO RECORD HIS TRAVELS, WHETHER FORCED OR FLEEING, ON HIS ZIPPO LIGHTER. LATER, HE WOULD retrace his journey, "On July 1st we were at Torigni, I don't find this on the map." H had difficulty pinpointing specific locations later because they were continually on the move.

July 2 & 3, 1944 found the recaptured POW's at St. Jores. "Clawson and I had been taken to a place we later learned was called Starvation Hill or Manor—following the escape. I am relatively certain that it was an observation post for an artillery command. There were a number of buildings there but we were primarily concerned with a single story building that was in the shape of a 'U.' The building was fronted by a courtyard, of sorts, with a low stone wall enclosure perhaps three feet high.

"Rear Echelon troops behaved quite differently from their counterparts on the front lines. They, apparently, used two techniques for interrogation. Clawson had been taken to one section of this U-shaped building and plied with coffee and cigarettes and politely questioned. I had been taken to the opposite side of the building where an English speaking German Officer, a Nazi Major, reinforced by a tough German Sergeant (a small non-com) treated me 'without gloves.'" Questioned for some length of time, H was beaten and slapped by the pint-sized non-com when he refused to answer their questions.

"I answered all questions with name, rank and serial number but after much persuasion, yelling and massaging with a rifle butt I lost my temper and referred to their parentage which did not help my cause. They could not believe that we had been part of the Invasion Forces and thought that we were part of a special mission so far back of the lines. At one point it appeared that I was to be the object of a firing squad."[222]

"When they found they couldn't get any information from him, they proceeded to line him up against a stone wall, the non-com drew back the regulation 10 paces. . . and then luck was with the lieutenant. For some reason which neither he, nor probably the Nazis themselves could fathom, the would-be executioner shrugged his shoulders and put his gun back in his holster."[223]

H continued: "Looking at the contour map of La Havre I have to believe that Starvation Hill was near St Jores, an obvious high point between Carentan and la Haye du Puits. At one time, while at Starvation Hill, the German major who had interrogated me led Clawson & I over to a grassy slope over-looking a vast valley through which ran a rail-road track. Where we were it was quiet and calm except for some 'dog-fights' between the Germans and American fighters. The Major mellowed and offered us sips from a jug of Aquavit. As we lounged there watching two or three of our planes were shot down. The Major gloated that his pilots had shot down 25 of our planes that day. Clawson informed him that: 'Shit, we made 150 of those planes [this] very day.'"[224]

On July 3, 1944 they finally left Starvation Hill/Manor. The two prisoners were escorted to Rennes where they rejoined others from their unit, the same one they'd escaped from 10 days before.

H and Clawson had simply circled the parameter.

JULY 1944

H and Clawson sat in the boxcar facing the secured doors. Recently recaptured, H was filled with a mixture of frustration, anger, and contempt that showed clearly on his face. It also tumbled out of his mouth when he was confronted by the serious Nazi guard, who did not speak English, but did read facial expressions.

The guard called for backup. H and two other surly sorts were hauled out of the car and walked over to a ridge where they were positioned to be shot. H thought he recognized one of his captors. The long, narrow face, the hawk nose...they called him Franz. Franz seemed to recognize him, too. The sour guard from his box car, however, appeared to take H's escape as a personal affront. H was certain the man wanted him dead.

He considered his situation. "It would be really rotten of you to shoot me without giving me a last cigarette," H exclaimed clearly, ever the smart alec.

His outburst seemed to amuse his captors, especially Franz. They laughed among themselves a moment, although his guard never cracked a smile.

Franz gave him a smoke. "Get back on the train," he directed him. H had avoided death again.

CHAPTER 34

TRAUMA OF CAPTURE

From the moment of capture, the American POW in German hands faced an uncertain fate. Aside from the daily deprivation of a prisoner, he shared the menace of Allied attacks on the enemy, ran real risk in attempted rescue and found danger lurking even in the final moments before liberation.
~ Military History, Vol. 1, No. 4 February 1985

"We arrived at Rennes and were shoved unceremoniously through the gates. We were met by an American Officer who informed us that the Senior American Officer wanted to see us immediately, and we were taken to the 'quarters' of Col. Paul Goode. Col. Goode questioned us quite extensively and then informed us that any further escapes we contemplated would have to include him," Harold recounted.

On July 4, 1944, they left Rennes by train crammed like sardines into box cars 40'x 8' in size. They were shipped like baggage to Schubin, Poland, then secured behind barbed-wire in a prisoner of war (POW) camp.

"The French boxcars were about half the size of the regulation U.S. variety, and the Germans put anywhere from 30 to 40 prisoners per car. It was a hot day in July when the group left Rennes, and the heat in the small cars was stifling. Lack of ventilation caused many to be overcome from heat prostration.

"The trip lasted 23 days and during that time the only food served them was one cup of water which definitely left none for washing purposes, a small portion of bread and sometimes a piece of cold lunch meat similar to bologna."[225]

Meanwhile, Marjorie received these telegrams:

---

8 July 1944

WESTERN UNION

=MRS MARJORIE H CARROLL=

THE SECRETARY OF WAR DESIRES MOT TO EXPRESS HIS DEP REGRET THAT YOUR HUSBAND SECOND LIEUTENANT HAROLD J CARROLL HAS BEEN REPORTED MISSING IN ACTION SINCE SIX JUNE IN FRANCE IF FURTHER DETAIL OR OTHER INFORMATIN ARE RECEIVED YOU WILL BE PROMPTLY NOTIFIES=

ULIO THE ADJUTANT GENERAL.

---

8 July 1944

WESTERN UNION

MRS H J CARROLL=

RECEIVED MESSAGE. KEEP UP HOPE

I FEEL HAROLDS ALRIGHT.

LOVE, ART

On July 14, 1944, H and Clawson finally received food from French Red Cross. They got off those cars to stretch their cramped limbs only once at a small French village near Tours. There the Germans allowed a French Red

Cross woman to give them soup that tasted like food from heaven after the diet they had been eating.

As they traveled in the hot and cramped box cars, hungry and dirty, they continued to be attacked repeatedly by bombs and machine-gun fire from the low-flying aircraft of both the Allies and Germans. There was straw on the floor, but after a short time, it was trampled to dust as the prisoners milled around in a desperate attempt to try to relieve their stiff muscles. Under the circumstances, it was difficult to rest, much less sleep.

Prisoner of War, Brooks Kleber, described his experiences as a Nazi POW to interviewer Alexander S. Cochran, Jr.:

Cochran – Somewhere along the line, you would have been transferred to the German POW system. At this point, officers and enlisted men would be separated, as would those in the air and ground forces.

Kleber – First we were sent to a Gulag, a transient camp, which is where the first formal interrogation took place, before being assigned to a permanent prisoner of war camp. Later on, I was amazed to find out that there, the air force prisoners were subjected to very intense questioning, confronted with their wives' names and their class at flight school. The Germans would overwhelm them with details, particularly their military past, which was very disarming and, I am sure, helpful to the Germans in information-getting.

I was placed in a boxcar with about 36 other officers. On July 4, 1944 I began a long 23 day train ride from Rennes to Chalons.

Cochran – That's a long time to be in a boxcar for what should have been a trip of less than 100 miles!

Kleber – We spent 10 days just sitting in a rail yard at Tours. Just prior to getting there, we were strafed by British night bombers. They hit only the engine, which was unusual as they normally attacked the whole train. One day we heard our own B-17s coming overhead. The German guards just locked the doors of our boxcar and headed for the bomb shelter. We watched helpless from the two little slits in the side of our boxcar and could see the bombs coming down.

Cochran – So you got strafed by the British and bombed by the Americans. That must have made you think about escape. Did anyone try to escape?

Kleber – About two or three nights out, we were able to get a plank up from the floor of our boxcar. As we were always moving very slowly, we were able to get six prisoners out on the bed of the car. When we stopped, they dropped off in the middle of the night. I don't know if any of them made it, as I knew none of them. But the next morning, after a headcount, the Germans discovered their disappearance. The senior American on the train was Colonel Paul Goode, "Pop," Goode.

Cochran – What did the Germans do to "Pop" Goode when they discovered the six missing prisoners?

Kleber – They took him out next to the train and dug a shallow grave. Next, they formed a firing squad, and "Pop" gave his West Point ring to one of his fellow officers. The Germans went to the last moment, but then placed everyone back on the train. It was very dramatic. [226]

Paratrooper with 503rd Parachute Emblem

LATER IN JULY 1944

There's not much to do in a hot, cramped boxcar but talk. And even that produces heat. The conversation was mostly between Dr. Gruenberg, Clawson, and H.

It started off a lively discussion about the varieties, the vagaries of heat. But after some time and quite a lot of sweat, the conversation switched to it being cold.

"Cold is cold," Gruenberg said.

H disagreed, "No. Cold isn't cold. There's a difference between Texas cold and Wyoming cold." He shuffled his feet in the straw thinly lining the box car's hot metal floor.

Gruenberg shook his head. "No. Cold is just cold."

"Not true." H was warming up to his topic. "Wind, blowing snow, and sleet all contribute to its being really much colder."

"Not possible!" Gruenberg stubbornly folded his arms across his chest. "Cold is cold."

"There are different types of cold. There's cold, and then there's cold," H emphasized with a shiver.

"You're crazy!" Gruenberg insisted.

"You've never experienced a Wyoming wind coming down from Canada," H persisted.

Gruenberg stood his ground, "Minus twenty degrees is minus twenty degrees!"

"Have you ever seen the snow?" H failed to get a reply to the last, finally ending the argument, although it resumed a few minutes later.

"So what about an ice cube?" Gruenberg asked. "Are you telling me the ice cube is colder in a glass of cold water than it would be in glass of, say hot tea?"

"Awwww!" someone nearby groaned. "Stop already!"

"Yes, I am," H answered. H and Gruenberg had to duck as they were pelted with straw and anything else available.

H went back to scribbling the date and the name of the town they'd just passed through on his Zippo lighter.

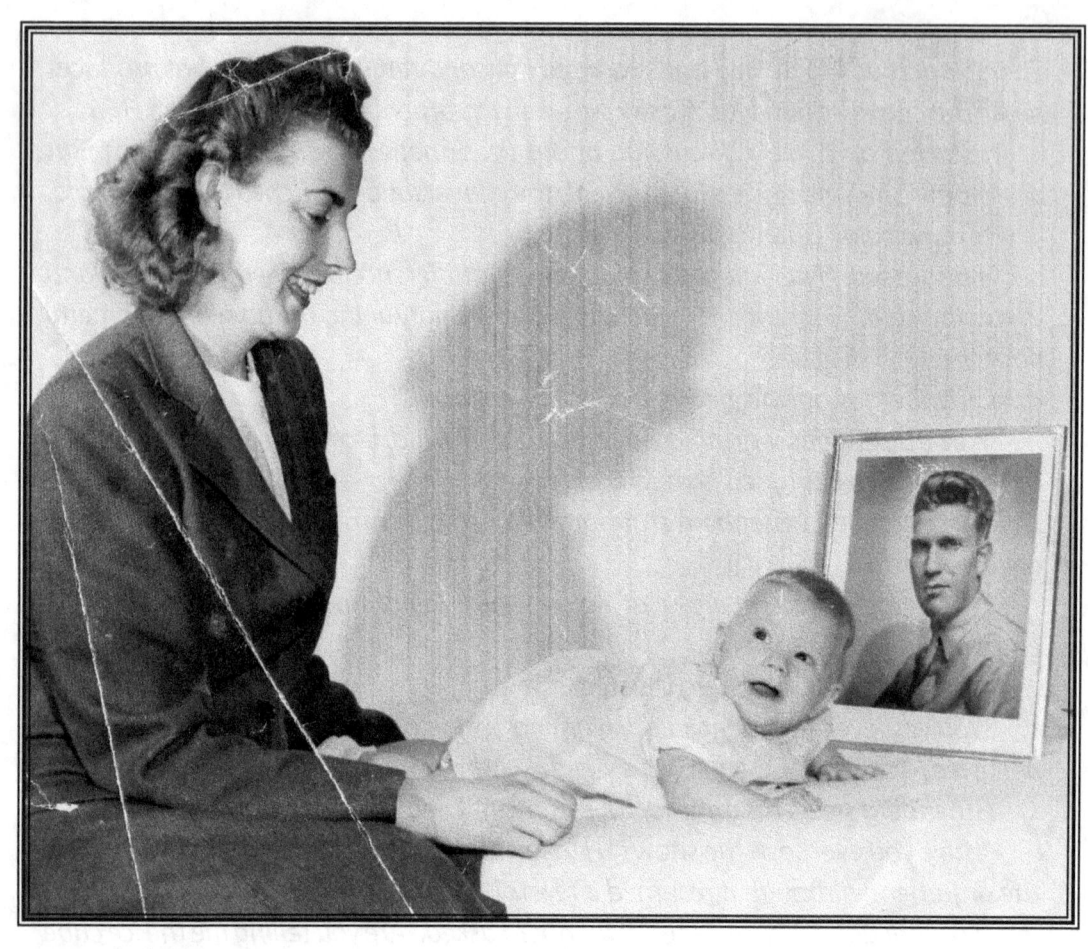

Marj & Pat

CHAPTER 35

THE ESCAPE BUG

The main reason the Germans had their permanent camps so far to the east was to be as distant from the battlefields as possible, thereby preventing escape. But this also placed them directly in the path of the Russian advance, which was moving faster than that of the Allies in France. ~ Brooks Kleber

H AND CLAWSON HAD TWO OBJECTIVES—FREEDOM AND RECONNECTING WITH THE MISSION. "ON JULY 16, 1944 we left Ville Dame for Nantes, Angers, and Tours. At Nevers, we escaped for three days. From there we marched on to Orleans. Me and Clawson and a few others escaped twice from the boxcars, but each time were recaptured and relined up to be shot."[228] They did not know if they'd been spared or if punishment was simply delayed.

They spent about one week en route to the prison camp at Chalon St. Marne. On July 28th they arrived, weary and wondering what was going to happen next. H found out when he was thrown unceremoniously and without explanation into the isolation cell, a six by six dirt hole in the ground with a slit in the door for air.

Just a few days later, on August 4, 1944, Anne Frank and her family were arrested by the Gestapo in Amsterdam, Holland.

H was given paper and pen to write a letter home. Marjorie was never far from his mind. He expected his mom was probably frantic and his father Arthur distraught. Of course, H didn't know that his father had passed away in June, just a few days after the invasion. He wondered daily about the whereabouts and welfare of his brothers.

Sadly, the delivery of H's letter was delayed. It didn't reach Myrtle and

Marjorie who still pined for news of him.

H pined for freedom, "on August 27, 1944, I was released from my isolation cell…and on August 29, 1944 we left Chalon Sur Marne for Bar Le Dev, Nancy, Aachen, Koblenz, then on to Limburg, Germany." In due course, H and Clawson arrived in Gepruft at the already infamous Stalag XII A.

H remained captive at Stalag XII A, located in Limberg, Germany, from September 1st until September 18th of 1944. Limberg POW camp was portrayed in Life magazine as a classic example of how the Germans starved their prisoners.

H continued to scratch his location, or where he believed he was, on his Zippo lighter. "On Sept 6, 1944, at Belfort, we escaped for five days." H noted. It was worth the effort and risk just to find something to eat.

H and Clawson were recaptured by the Nazis at gunpoint and forced back to the miserable Stalag XII.

"On Sept 11, 1944 we left Limberg for Kassel-Noadhausen and Frankfort-Poznar arriving around Sept 18, 1944."[229] H and Clawson arrived weakened and half-starved, and took up residence from September 18, 1944 until January 28, 1945 at Poland's notorious Oflag 64.

---

18 Aug 1944

Prisoner of War Camp

11872 U.S. Censor

Stalag XII A

I have been taken prisoner of war in Germany. I am in good health—slightly wounded

We will be transported from here to another Camp within the next few days.

Please don't write until I give new address.

Kindest regards

Christian Name and Surname: Harold Carroll

Rank: 2nd LT.

Detachment:

Darling:

No doubt by now you have been notified that I am 'missing in action'—I'm sorry you were worried this way—really fairly healthy and whole. I have been a prisoner since D-Day. The war seems rather remote to us here; for the most part it is peaceful and quiet; as matter of fact today you can hear an organ playing in some chapel in the Stalag.

This is not a permanent camp but a transitional Stalag so only this letter may be sent and none received, perhaps soon though you may write me through the Red Cross. I am very anxious about you. Please take care of your self.

(sentence blacked out) you might notify the Sheridan paper about me so others will know.

We have received a Red Cross parcel recently and the future begins to look bright.

Hope I shall be able to write again soon. Contact the Red Cross for my address—Don't worry.

Love, Harold

AUGUST 1944

H and Clawson escaped in the wee hours of the night. Keeping their bodies low to the ground and staying in shadow they made their way stealthily beyond the camp's sentries. Heading for the woods, they found themselves maneuvering through a field of sleeping Nazi soldiers. H stepped ever so carefully over a body, which while prostrate, still clung in slumber to the carbine at his side, finger twitching. H's heart raced, adrenalin surged through his body and kept him alert and focused, if jumpy.

The hedge rows dividing the fields of wheat and barley were so thick you could not see through to the other side. H stopped to relieve himself on one side of the shrubbery. Just as he was in a most vulnerable position, he heard a rustling sound. A German officer on the other side of the hedge row had approached for the same purpose. Hearing the sound of urination just beyond his line of vision, the German said nonchalantly, "Was ist los?" (Roughly translated as, "What's going on?")

There was a slight pause and only the sound of trickling water could be heard. Then H responded, "Los ist los." (Generally, this phrase is meaningless in German.)

The German turned aggressive and hurriedly attempted to get through the sturdy green wall to H. He lost his life in the attempt. H shot him—H had never shot anyone before, not up close and personal. Returning fire at dark, undistinguishable moving forms that were shooting at him when he landed on D-Day was instinctive self-defense. Having a conversation with someone before you fatally shoot them felt different. The situation stunned Harold and he couldn't shake it. Clawson hurried him along. "Come on, Carroll, keep low."

A few days before D-day. From left: Harold Miller, Charles Christian, Max Domina, Stanley Weinberg, Harold Carroll, and Carl Clawson.

H's POW ID from Oflag 64

CHAPTER 36

OFLAG 64
Thin, green gruel. ~ H Carroll

OFLAG 64 WAS A PRISONER OF WAR CAMP FOR U.S. COMBAT OFFICERS. SHORT FOR OFFIZIERLAGER, "OFFICERS CAMP," it occupied the grounds of an old college campus in northern Poland. It included a large, three-story stone building, three barracks, a sports field, chapel, infirmary, canteen, and huts for classes. Some POWs had been there so long they'd devised their own college curriculum.

The prisoners, or "Kriegies" as they called themselves, (short for Kriegsgefangenner), many who had been there since 1943, had to organize their own activities to keep from going stir crazy. Their activities included amateur theater and a hand-written newspaper.

And they were being slowly and systematically starved to death. "The limited menu was as bad as that of Limburg," H remembered. "The only highlight in the drab and dreary meal was the three boiled potatoes they were issued once a week. The rest of the time, the menu varied not at all, consisting of a ration of bread (about one-fifth of a loaf), a thin, green gruel made from sugar beet tops boiled in water, and varied occasionally by the same type of stuff made from cabbage leaves. Once a week also, they were allowed one dab of margarine to smear on the hard bread ration," H told Deryk Wills when interviewed about the camp.[230]

"We newcomers received at most two Red Cross Parcels from Sept 18 until we left. It is possible that my memory does not serve me right and there could have been 1 or 2 more Parcels—but I don't think so. Naturally, on such a diet, the prisoners lost weight."[231]

The Red Cross parcels were supposed to be just a supplement. But the nourishment the Nazis provided was less than subsistence. And although the parcels had been distributed every week at first, soon they were rarely seen and were hoarded by desperate men when distributed. They contained such necessities as powdered milk, instant coffee, a chocolate bar, and a tin of meat. The Germans blamed the low supply on the Allies; they said trains from neutral Sweden and Switzerland were being bombed and so the Red Cross parcels couldn't get through. The Kriegies became obsessed with thoughts of food.[232]

Prisoner of War, Brooks Kleber, hunger expert, described his experiences as a Nazi POW to interviewer Alexander S. Cochran, Jr.:

Kleber – Right after I got there, the Red Cross parcels for prisoners of war stopped, and I never got any mail from home. Now, these guys had been there for more than a year, receiving all kinds of stuff from home, as well as the weekly Red Cross parcels. There was a lot of stuff that you could save from them. When it became obvious in December that the Russian advance was going to force the evacuation of the camp, you might have expected some sharing of these surpluses. But there was none. Later on, when we left the camp, these guys made improvised sleds full of things they had saved. And they would discard it rather than give it to us. I remember the older guys using margarine for lamps, while I would have loved to use it with my bread.

Cochran – Earlier we mentioned a "shadow staff" in permanent camps. Did Oflag 64 have one?

Kleber – Yes. It was just like any other Army staff, with a commander, a chief of staff, an S-1, S-2, S-3, and S-4. They imposed and supervised the internal order and direction of the camp. They were very important in the overall functioning of the camp. Remember, after we were captured, the only Germans we saw were those guard companies. We took care of ourselves.

Cochran – You mentioned that the American leadership at Oflag 64 had ordered no more escape attempts.

Kleber – yes, there had been escapes, but with the war "so close" to being over, the leadership felt this would just waste lives. As I remember, we agreed with this approach.

Cochran – I guess we now forget that most people felt the war would be over in late 1944—until the Battle of the Bulge and the fierce defense by the Germans of German home soil. We read about German spies in prison camps. Did you have any problem with this?

Kleber – The prisoners had their own system to deal with this. First of all, you normally knew someone in the camp. But if someone came in that knew no one, he was very subtly interrogated to make sure that he was not a German plant. You tried to find out where he was from and get into those areas that he would not have been briefed on in his cover story.[233]

September 21, 1944   **POW**

Darling:

I am in good health now. Only wish I could be with you. We have a good library and sports equipment—beer and cigarettes. Don't worry about me.

    Love,

    Harold

JAMES J. KOCH
PRESENTS

"the MAN WHO CAME TO DINNER"

By
HART and KAUFMAN

OCTOBER 11th, 12th, 13th, 14th, 15th — 1944

LITTLE THEATER
OFLAG 64

OCTOBER 1944

Loud laughter pierced the silence of the camp. A group of Kriegies stumbled into the glare of the floodlights that swept methodically across the barracks and outbuildings. One drunken man shoved another and angry words were exchanged, followed by shouting. A tall, lanky POW took a swig from a bottle and passed it to a shorter, laughing comrade. They stood to the side, watching the fight develop and talking about it in loud, sarcastic voices, egging it on. A fifth man attempted half-heartedly to break up the brawlers.

"C'mon you guys, cut it out!" he shouted.

They were making so much noise that formerly sleeping men started pouring out of the barracks, rubbing their eyes and elbowing each other. A melee was about to ensue.

The plan was simple but genius. The "drunks" caused such a ruckus in the middle of the night that they were removed from camp and brought to a small jailhouse on the other side of the menacing fence, the free side.

Before they reached the jail, they overcame their captors and slipped into the black night. They had escaped the dreaded machine-guns and barbed wire. But their freedom was short-lived. They were soon recaptured by armed guards and sent back to Oflag 64. They didn't suffer much more than an ego bruising, but their recommendations for future escape attempts were ignored.

October 1, 1944   **POW**

Darling:

You know, it certainly seems like the weeks are rolling by fast – but not fast enough. Hope you've heard from me by now. When you write me tell me about Bob. I haven't had a bit of news of him. Wish you'd contact the Red Cross and see what you can do about sending packages; There are things that I need badly like razor blades, vitamin pills, socks, underwear, pipe and tobacco and cigarettes. Wish you could send an over-coat or blankets—I have but very little clothing and it's plenty cold here.

Certainly hope you are feeling well now Darling, how about sending news and pictures of yourself and sport. Gosh but I'd like to see you! Well, I hardly know what I can say. Guess I'll slip over and see Callopy tonight.

Take care of yourself and don't worry.

Lots of Love,

Harold

November 4 1944

WESTERN UNION

MRS MARJORIE H CARROLL=

REPORT JUST RECEIVED THROUGH THE INTERNATIONAL RED CROSS STATES THAT YOUR HUSBAND SECOND LIEUTENANT HAROLD J CARROLL IS A PRISONER OF WAR OF THE GERMAN GOVERNMENT LETTER OF INFORMATION FOLLOWS FROM PROVOST MARSHAL GENERAL=

November 9, 1944 **POW**

Darling:

Feeling well—anxious for a letter from you; hope you are receiving my letters. Hope you and Sport are well. Don't worry—just write very often and send what packages you can, they are much needed.

    Love,

    Harold

November 20, 1944   **POW**

My Darling:

You know, we're certainly going to have a lot of living to catch up on when I get home. Sure hope I hear from you soon. Bitter winter here now—lots of snow and I can't even enjoy it as I usually do. Just a bit hungry. Are the packages on the way? How's Sport? Hurry and tell me about him, Honey. Are you feeling well? Be thinking of a place to spend a month or so sick leave when I get home, brush up on your cooking and don't forget the apple pies. Wonder if you've received my belongings from the Regt. Yet—hope everything's there—could use some of the toilet articles—specially razor blades. Have Art write me will you—and Lucy and your Mom—Give all my love. Be happy—lots of Love.

    Harold

November 26, 1944   **POW**

Darling:

Well, another Sunday and still no word from you—It gets no warmer and we get hungrier. Sure praying for a food parcel from you—even the Red Cross has let us down. No parcels from them. Could use a few cartons of cigarettes and tobacco, too. How are you, Honey? Should be pretty much your old self now. Sure anxious to hear about Sport. Don't let the Red Cross tell you not to send food. Forgive me if I keep mentioning it but it's necessary. What's new back there—Tell me about everything. Store up a few choice bottles of Scotch and Bourbon won't you, along with the apple pies. It's hard to write now and this Polish weather does nothing to help. Write.

       Love,

       Harold

November 30, 1944

WESTERN UNION

MRS. MARJORIE H. CARROLL

FOLLOWING ENEMY PROPAGANDA BROADCAST FROM GERMANY HAS BEEN INTERCEPTED QUOTE AM WELL AND HAPPY AT OFLAG 64 GERMANY

YOU MAY WRITE TO ME HERE SECOND LIEUTENANT HAROLD J. CARROLL 01308095   UNQUOTE   THIS BROADCAST SUPPLEMTS ALL PREVIOUS REPORTS

LERCH, PROVOST MARSHALL GENERAL

December 8, 1944

Dear Art,

I wrote you a few weeks ago but before the letter was posted the dog proceeded to chew it up. Let's hope I get this off before he gets it.

Your Mom has gone back home now. Guess she thought she had better be there to see about the house. We had a nice visit while she was here and she had a wonderful time playing with Pat. Hope you received the pictures she enclosed. We took some more a couple weeks ago if they are at all good I'll keep sending them. Pat is undoubtedly going to be a little red head she is displaying a bit of the proverbial red heads temperament now. As a rule she is very good and so little trouble — do wish Harold could see her. I'm certainly glad to hear your wonderful news I think it's wonderful, a niece, (or nephew)? my first! I received a letter from Milly yesterday. We thought perhaps if she were going home for X-mas she could stop here en route and she and your mother and I could be together for awhile. I do think she is right in not traveling now, however, and shall perhaps be able to get together with her before too long. I'm so anxious to meet her.

Have sent Harold's first POW package on to him plus a tobacco package. I do hope they reach him in good time. I presume that your mother has written to you that I received a German propaganda radio message from Harold saying he was well and happy at Oflag 64, and for me to write. So far I have received 30 cards and letters from short wave listeners relaying the message. Now I'm waiting for a letter from him.

X-mas is practically upon us again. I'm not looking forward to it much but am setting my cap for a really festive next year with you and Harold both back with your families. Here's hoping I'm not being overly optimistic but it does sound very possible to me.

Hope everything is ship shape for you and have a merry x-mas with all the trimmings.

  Love,

  Marj

December 18, 1944   **POW**

Darling:

   One week till X-mas. Hope I hear from you by then. Everything is much as usual here, well but not happy. How are you Honey? Tell me about the baby and don't forget the photos. Playing a lot of bridge and reading a great deal—That's about all there is to do. It's been very cold here—some snow—doesn't get below zero yet. I know there's not much to these letters—only want you to hear from me. Keep packages and letter coming and don't worry.

   Lots of Love,

   Harold

OFLAG 64 STAFF

Col. P. R. GOODE, Senior American Officer
Col. G. U. MILLETT, Jr. Executive Officer
Col. F. W. DRURY

Lt. Col. J. K. Waters   Lt. Col. L. Gershenov
Lt. Col. J. D. Alger   Major K. R. Hanser
Lt. Col. W. J. Martz   Capt. F. M. Burgeson
Chaplain C. F. Glennon

BARRACKS COMMANDERS

Lt. Col. G. J. Barson   Lt. Col. Ch. Jones Jr.
Lt. Col. E. P. Cummings   Lt. Col. C. W. Kouns
Lt. Col. M. H. Cooler   Lt. Col. R. S. Palmer
Lt. Col. E. C. Hardaway   Lt. Col. J. F. Skells
Lt. Col. R. J. Herte   Captain M. C. Smith
Lt. Col. W. G. Hopkins   Lt. Col. H. W. Sweeting
Lt. Col. D. R. Yardley

CHRISTMAS 1944

Druck: Willi Kricks, Altburgund

December 23, 1944   **POW**

Darling:

   You're very much on my mind today—x-mas is here and for one more year we must be apart. Our make-believe X-mas here has very little meaning, needless to say, and I'm only thankful that I'm alive and have a person like you waiting for me. I want you to be very happy, Darling, and don't worry about me—only pray that we may soon be together. Please give my best to everyone. Spoil the baby for me and have fun. Have a lovely x-mas, Honey, and pray that this next will be our happy year.

   All of my Love,

   Harold

Darling:

It's the last day of the year—didn't see much of it did we, Honey? A thoroughly bad year but it gave us Pat and that makes it very wonderful. Gosh, but I'm thrilled about Pat! Recd. Your letter Friday, my first mail, and have been walking on top since. Please write very often, Darling, and tell me all about Pat and you. She must be an image of you—and that's really bringing beauty into this old world. I'll bet she is a great help to you—save the spoiling of her for me. I didn't know of Art's wife, you know—have them write—Golly—old Art! Patricia Lynn—it's really wonderful, Darling, wish I could tell you how happy it makes me. Really a bright light in 'Kriegie' life. Write very often, Darling. Give my love and blessings to Milly—have her write. My love to everyone there. Take good care of yourself and Pat and pray I'll be home soon. Write.

        All my love,
        Harold

KRIEGIE HUMOR
It saved us when the going got tough

1. "You jump first, Joe, I don't want to be a hero."

2. "Yow! This can't happen to *me*." (That's what every Kriegie thought.)

3. "Hope they didn't see me. Now all I got to do is cross Germany, France and the Channel and I'll be back in good old England."

4. "For you der var ist ovfer."

5. "Think of our poor women and children." (That was supposed to soften us up.)

6. "Aw, have a heart, stomach, this is my day to think about women!"

7. "... and when good Kriegies die, they go to heaven, where there is an easy chair, slippers, plenty of smokes, a radio, and apple pie and milk."

By Sgt. Lester H. Russell

H Carroll, left

CHAPTER 37

NO ESCAPE

*The German Camp Commander, Oberst Schneider, was "a portly old man...
all he needed to resemble a Hollywood villain was a monocle. He would stand
before us with his enlisted interpreter beside him; then he would say a few
sentences, stop, and take one big step to the left. The interpreter would tell us
in English what the Camp Commander had said. This was followed by
more of the same, always with that big step to the left.
A long speech could result in a 30-foot progression to the left."*
~ Oflag 64, Fifthieth Anniversary Book

OFLAG 64 OPENED WITH ONLY ABOUT 150 OFFICERS, BUT BY THE TIME H AND CLAWSON ARRIVED THERE WERE OVER 1,400 men. The average age was twenty-seven. The soldiers had a lot of time on their hands and channeled their restlessness into various activities including reading books from the camp library, working in the garden or the repair shop, or just playing cards. H and Clawson continued with their favorite pastime—escape planning. Unfortunately, Oflag 64 had many armed guards watching from machine-gun towers and was surrounded by barbed wire.

"All escape plans had to be approved by the Escape Committee, who considered if it could endanger other, already approved attempts in the planning stage and also, if discovered, if it could bring down a hail of bullets, endangering the rest of the camp. The man who originated an approved scheme had 'ownership' and could name as partners only those needed as a minimum to make the plan work. Once an escape plan was approved, it became a camp-sponsored project. Any POW who briefly tasted freedom and was recaptured went to the bottom of the 'waiting list.'"[235]

The prisoners at Oflag 64 had an elaborate tunnel digging plan: "Diggers sent the dirt back up in old Red Cross boxes which were then stored in the low attics of the barracks until they threatened to break through the ceilings, which happened on one occasion. Then the clever tunnelers sewed sacks inside the trousers of some of the kriegies and filled them with dirt. As the kriegy walked around the 10-acre camp, he would pull a string releasing small quantities of dirt from time to time...Meanwhile, an almost identical tunnel was being dug by the American and British flyboys at Stalag Luft 3, a huge compound also located in northern Poland. This one was completed first and was the scene of the 'Great Escape' of motion picture fame. It was, unfortunately, a great disaster...the exit came up a bit too close to the German guard barrack, so that only three men actually got away, all of the rest being recaptured within a few hours, and 50 of them shot upon Hitler's orders. These murders convinced our Escape Committee to close down the Oflag tunnel project."[236]

H was somewhat impressed with the prisoners at Oflag 64. He remembered it as a tightly run camp, especially compared to the others he'd experienced up until then. "The Oflag obviously was well organized. I was not involved in the tunnel digging, or in any of the 'organized planning' of the Senior Staff," H told Clarence Meltesen when interviewed about the camp.[237]

H and Clawson had gained considerable experience in escape planning and execution. They also had plentiful experience being recaptured. At Oflag 64 they found themselves uninvited to the planning party and on the fringe of a long-established group of POW's. They were newcomers.

"I would have to express a certain amount of disdain as regards the 'class system' in the Oflag. I was, probably, somewhat envious of those inhabitants of the 'Big House,'" H admitted.

"Was Oflag 64 Disciplined? Yes. It was a period that, if we had had our way, Clawson and I would have made another break. But, as you know, we had to submit a plan—which would have been denied—so, I guess that we just sat and vegetated. I played football, read some, walked a lot. I had managed to keep with me a miniature deck of cards and found certain 'popularity' with some Kriegies who wanted them to play Bridge. Naturally, I went with the cards."[238]

H and Clawson bided their time at the Oflag. But they were too restless to be still for long and change was in the air.

Back home, Marjorie attended a regular card game with other army wives. The telegram arrived while the ladies were playing Bridge. Marj hurried to open it.

Marjorie had recently listened to a German propaganda radio message just a few nights before. The message was clearly stated: "Write to H." And so she did. She also wrote to his brother Art.

Radio waves, like letters, had become a life-line between soldiers and loved ones back home. People who listened to radio short waves became skilled at hearing important details and then passed the information on to help people like Marjorie who were desperate for information. Some became well known, like "Axis Sally" who broadcast information about German POWs.

Short wave listeners "sat in dimly lit rooms far into the wee hours of morning. Their faces illuminated by the faint, yellow glow of a short-wave radio dial, they carefully tuned through crackling frequencies during the dark days of World War II to listen to the propaganda broadcasts from Berlin, the enemy capital nearly 4,000 miles away. They were everyday citizens who independently served as ex-officio intelligence agents, sometimes finding hope and joy for the families of American and Canadian POWs in the enemy broadcasts designed to create only despair and disillusionment.

"The broadcasts included names of POWs, sometimes service numbers, hometowns, and the names of family members." Thanks to the short-wave radio listeners information was passed along that might have saved many a family member from another long night without sleep.[239]

H continued to write home in spite of his cold, numb fingers and the lack of any response.

Prisoner of War, Brooks Kleber, hunger expert, described his experiences as a Nazi POW to interviewer Alexander S. Cochran, Jr.:

Cochran – How about news of the war? How did you know what was going on? You must have been hearing something to be so discouraged in the winter.

Kleber – We never learned any true news from the Germans unless it was bad news. The Bulge – we got an accurate treatment of this, you may be sure. The rest of war news came from our secret camp radio and the BBC. This was in turn distributed by word of mouth in pyramidal fashion. The code word was, "The bird is going to sing," and we would gather in small groups to learn the latest from the war front.

Cochran – The combination of news from the BBC, plus reports on the Bulge, must have had a disastrous impact upon camp morale.

Kleber – Morale got lower and lower. The Red Cross parcels stopped. Weather got colder. The guards told us that the Russians were getting closer. Sometimes we thought we heard fighting. There was a definite feeling of uncertainty. We sort of sensed the camp was going to be closed down soon.

Cochran – And then one day in January you were told to prepare for a 10-kilometer march to the railroad station and another train ride to a new camp.

Kleber – Yes, but this time there was no train. For the next 43 days, we marched more than 300 miles, away from the Russian advance and Poland, back to Germany. We marched during the day and slept at night in large barns, ones that would hold more than 1,000 prisoners.

To all Prisoners of War!

The escape from prison camps is no longer a sport!
Germany has always kept to the Hague Convention and only punished recaptured prisoners of war with minor disciplinary punishment. Germany will still maintain these principles of international law. But England has besides fighting at the front in an honest manner instituted an illegal warfare in non combat zones in the form of gangster commandos, terror bandits and sabotage troops even up to the frontiers of Germany. They say in captured secret and confidential English military pamphlet

THE HANDBOOK OF MODERN IRREGULAR WARFARE:

"…the days when we could practice the rules of sportsmanship are over. For the time being, every soldier must be a potential gangster and must be prepared to adopt their methods whenever necessary."
"The sphere of operations should always include the enemy's own country, and occupied territory, and in certain circumstances, such neutral countries as he is using as source of supply."
England has with these instructions opened up a non military form of gangster war! Germany is determined to safeguard her homeland, and especially her war industry and provisional centres for the fighting fronts. Therefore it has become necessary to create strictly forbidden zones, called death zones, in which all unauthorized trespassers will be immediately shot on sight.
Escaping prisoners of war, entering such death zones, will certainly lose their lives. They are there fore in constant danger of being mistaken for enemy agents or sabotage groups.
Urgent warning is given against making future escapes!
In plain English: Stay in the camp where you will be safe! Breaking out of it is now a damned dangerous act.
The chances of preserving your life are almost nil!
All police and military guards have been given the most strict orders to shoot on sight all suspected persons.
Escaping from prison camps has ceased to be a sport![240]

Letter to My Baby Back Home, a few days before D-day, Photographer: Harold Carroll

CHAPTER 38

LETTER FROM HOME

The code word was, "The bird is going to sing," and we would gather in small groups to learn the latest from the war front. ~ Brooks Kleber [241]

STUBBORNNESS CAN BE A POSITIVE TRAIT. TENACITY OF SPIRIT KEPT H WHOLE. BUT IN ADDITION TO STARVATION, deprivation, and lack of cleanliness, H was delusional with fever. He was hospitalized from October 15 to November 5 for treatment of festering flesh wounds that had never been previously treated. He also had the misery of bleeding constantly from "scabies." When he was finally able to write home, he wrote regularly. His letters avoided talk of his condition. H didn't want to worry Marjorie unnecessarily. All letters were censored by the Nazis, anyway. Despite all of his writing, he did not receive a letter in return until after Christmas, 1944.

"On December 31, 1944, I received my first letter from home." H had received only one letter mailed months before. When the Russians rolled in, they found a room stacked high with prisoner mail the Nazis had neglected to issue. In that stack were three more letters for the lieutenant, one of which told him of the birth of a daughter in Cedar Rapids, IA.

Many letters never reached their intended destination. Some disappeared leaving no trace. Some letters were, sadly, returned with the writer left to turn the unopened envelope over and over in his hands and glumly wonder why.

January 8, 1945   RETURNED TO SENDER by Direction of the War Department, Undeliverable as Addressed.

2nd Lt. Harold J. Carroll

Prisoner of War No. 83959

Oflag 64

Germany

Hi My Darling

I am now enroute on what will probably be my last vacation in some years. I'm on my way to N.Y. to see Milly. I'm sure from all reports that she is a swell person and I'm looking forward to meeting her. Lois is taking care of Pat while I am gone; However, don't plan to stay long for I've been gone but a few hours and am already lonesome for her. She's growing so fast and is such an excellent and happy baby. She tries to sit up now and giggles and goo's a lot. She had a long talk with her Daddy's picture yesterday. Wish it could have been with her Daddy.

I'm here now and so far think it is wonderful. Have play tickets and really expect to see the town and above all, rest!

Milly is a perfectly swell person and evidently an excellent artist, such a tiny person!

So far I think the city is wonderful but no place for Pat at the present at least.

I miss you so much, Darling, and think of you always. I do love you so very much, Honey, only wish you could be with me.

Your Marj

JANUARY 1945, NEW YORK CITY

Marjorie sat relaxed and day-dreamy on the train bound for Iowa. She was returning home from New York City. It had been her first trip to the east coast. She felt more sophisticated and sat taller, having spent almost three weeks in the "Big Apple." She was dressed in a charcoal jacket and skirt made of light wool, tailored and flattering. She'd slimmed right up after the birth of Patricia and looked elegant and sleek in her New York clothes.

"I invited her to come and visit NYC," Mildred remembered. "And her Aunt Hazel and mother, Merle Harris, thought it was a good idea and helped her out. Little Patty stayed with Marj's sister, Lucy, who was also living with grandma and auntie at the time."

Winter wind and sleet pelted the roof and battered the windows, but the train was warm and cozy. Her visit with Milly had been splendid. The most wonderful experience of her life so far, next to meeting Harold and having a baby, of course. The excitement of the city took her breath away.

Marjorie had arrived at a Manhattan apartment filled with four lively and lovely young women. Mildred was such a wonderful artist and had such wonderful friends, Marj thought. And she seemed so relaxed and at home right there in the heart of it all. Marj was as much in awe of sophisticated, modern Milly as she was of where she lived.

"We did have fun! We always had show tickets because one of the girls worked for Better Homes and Gardens and did reviews of any new restaurant or any new Broadway show that opened. We would not have been able to afford to go otherwise," Milly admitted.

The other girls drank highballs but Mildred was a teetotaler. "I was the only one who didn't drink alcohol. I was drinking coke that night. But the next morning I was the one who had the headache!" Milly and her roommates had thrown a party for small-town Marjorie, an experience she'd never forget.

Marj traveled on a train with comfortable berths, private washrooms, and a luxurious dining car. They'd long since left the city behind and were rumbling through Ohio cornfields when Marj made her way to dinner.

She'd had a glass of gin and tonic before dinner. The chef was skilled, and she enjoyed her meal with a glass of wine. The rolling of the train with its rhythmic rumblings, the tinkling glasses and murmuring conversations of those at nearby tables, lulled her into a bit of a trance.

She stared out of the window watching rain pelt the glass and then slide into patterns. Scenery went by in a blur. She thought of her husband, wondering if he were well this cold evening far on the other side of the world.

Then she knew. Harold had escaped from the Russians. It was like a light bulb had gone off in her head. She just suddenly knew it. She said it aloud, "Harold has escaped from the Russians!" And again, louder, as she stood telling everyone, "He's escaped! Harold's escaped from the Russians!"

Marjorie's premonition was correct. On February 29, 1945, H escaped from the Russians...

```
   January 9, 1945   RETURNED TO SENDER by
Direction of the War Department, Undeliverable as
Addressed.
   2nd Lt. Harold J. Carroll
   Prisoner of War No. 83959
   Oflag 64
   Germany
```

Hello Honey,

How's my very favorite person? I'm still visiting Milly and having a real good time. Saw a couple good shows and some of the town. I'm getting pretty anxious to get home to Pat so shall save lots of good things to see when we can see it together.

Did I tell you that our Pat is going to be a blond after all? I guess her red hair has come and gone. I'm buying materials here so when I get home I can really start sewing for her. Am going to a sale as soon as I finish this. I'm trying to collect things for that vine covered cottage.

I do hope you're able to keep busy. I sent another pair of glasses on in your last box in case you need them. Wish I could do so much more for you, Darling, but some day I shall be able to. Please remember and know how very much I love you, Darling, and that you are always here with me. Take good care of yourself, Darling, and don't worry about Pat and I. We are taking good care of ourselves until our Daddy comes home to do it for us.

All of my love and a big kiss from us both, Marj

```
   February 18, 1945 RETURNED TO SENDER by
Direction of the War Department, Undeliverable as
Addressed.
   2nd Lt. Harold J. Carroll
   Prisoner of War No. 83959
   Oflag 64
   Germany
```

Dear Son:

I haven't heard from you for some time now but know you have been on the move. Doubt very much if any of our letters have reached you, but will keep writing. And hope it won't be so long now until we will hear something quite definite from you. Marjorie sent boxes of food to you at different times after hearing from you but if our letters don't reach you it is doubtful if anything else will.

Marjorie writes that little Patricia is growing so fast and is getting so cute. I am anxious to see her again and do hope you will get back while she is still tiny. Marjorie says she looks more like you every day, and is so sweet and good. She is really a darling baby and you will surely enjoy her. Do hope you got the pictures Marjorie sent of her.

Our winter continues on and on here with snow a foot deep everywhere and pretty cold every night. It was five below zero when I went to work yesterday morning and it didn't warm up much all during the day. The days are getting longer now and I am hoping spring isn't too far away. But how we are all hoping it won't be long till you will be back home again, maybe back in time to do some fishing this summer. I imagine it has been some time since you had a fishing rod in your hand. Your things are still sitting around pretty much as you left them. I go down and clean up your den. I have been pretty well this winter and keep awfully busy every minute, it helps to pass away the time if you can keep occupied. I hope we will hear from you soon and that you are well. We are all thinking of you here and pray always that God will take care of you and bring you back safely before so long…

With all my love,
Mother

A few days before D-day, photographer: Harold Carroll

CHAPTER 39

THE FIFTH AND FINAL ESCAPE, SORT OF
*We escaped—just the two of us — from the Nazis
while they were hastily shooing the prisoners eastward
with the huge Red army drive in full steam right on their rear flank.
~ H Carroll*

Clawson, H, and the other POW's stayed in Oflag 64 until their Nazi captors discovered the close proximity of the Russians troops and prepared to flee the "red steamroller." The Red Steamroller referred to the colossal Russian army moving intractably forward towards them.

The German guards told the POWs to get ready to move. On January 21, 1945, they began their long, cold and hungry walk to Exin.

H and Clawson plodded along with the rest of the officers who were deemed to be physically capable. About 40 in the camp had to be left behind in the prison; they were too ill to make the journey.

"There was good reason, too, along the Nazi line of thinking," H noted, "for by that time, the Russians were in full swing across the country and were getting somewhere close in that neighborhood. I then escaped from the Germans when they attempted to move the prisoners in Oflag 64 into Germany."

H and Clawson and the other POWs were a forlorn looking group. They'd layered any piece of clothing they could get a hold of over their tattered uniforms to keep warm. Huddled close together in line for shared warmth, they walked through the snow and freezing rain over the rough terrain.

H reached for his canteen with numb, chapped fingers, but the water in it was frozen solid. He kept putting one foot in front of the other, following the soldier in front of him, both of them thirsty, hungry and tired.

"We walked in the column one-day (20th), and part of the 2nd day (21st) to near Kcynia." The first night out they bedded down in the hay at a Polish farm, and the escape 'bee' hit H and Clawson at the same time. "We, Clawson & I, knew damned well we were not going to be herded into Germany in that miserable weather." Many of the other prisoners simply buried deeper in the hay the next morning when the Germans began the task of rounding them up, but H and Clawson had a better idea.

"'We escaped—just the two of us"—from the Nazis while they were hastily shooing the prisoners eastward with the huge Red army drive in full steam right on their rear flank. They dodged into some bushes by the road as they marched along; kept out of sight during the day, and that night, circled back to the Polish farm.[246]

"After our escape we hid out in a large Polish home, a very fancy building, near Kcynia (Gernickie), until February 1st. We then started out for Znin (Exin) taking 2 days (2nd & 3rd)."[247]

On Jan 26, 1945, the "Red Steamroller" entered Poland and Soviet troops liberated Auschwitz. The Allies were coming.

On their way to Exin, H and Clawson traveled through the cities of Kutno, Biolystok, Grotno, Lublin, and Vladimir. H recounted:

> On Feb 1, 1945 in Znin, at a Rail Road station we boarded a one car train and traveled over-night to Wrzesnia (Wreanchin.) And from Wrzesnia we traveled east on a main road, partially, to a small town, I think, Chelmno, where town officials greeted us, welcomed and fed us then we proceeded back to the main road.
>
> On February 3rd and 4th we hitched a ride on an American truck and rode until dark, pulling into a Polish farmhouse at dark and here we spent the night.
>
> Shortly afterward, the Red army, and what an army it was pushing bumper to bumper along the road, came in sight and as a routine gesture, an officer came into the farm house to see if any Germans were hiding. One of the prisoners, who had made his escape via the hay, wasn't sure which nationality the Russian was and said 'Sprechen Dutch' or words to that effect in German.
>
> Neither could the Russian tell who the prisoners were; their conglomeration of clothing gave no hint, and when the Yank prisoner spoke in German, he immediately pulled a hand grenade out of his

pocket and made ready to blast the group into bits. Before he could raise his arm to send the grenade on its mission for death, one of the prisoners yelled, "Americansky! Americansky!"

That settled the identification problem at once, but just for a second, death had been close. The Russian and the Yanks shook hands, pounded each other on the back, and immediately began to get acquainted.

On February 5th, we walked, hitched rides on Russian driven trucks, caught a ride on a train and walked a lot until we finally reached Warsaw. We proceeded on the next morning, in the same truck, to Kutno and then into Praga, and from Praga to Rembertow.[244]

H walked away from firing squads, and other close calls with death, and later he mostly kept to himself the unspeakable things he'd experienced as a soldier. But he was a witness. "In Praga on February 5, 1945, we talked with an English-speaking Polish 'Gentle-woman' who told us of the gas chambers in Praga, and showed them to us, then directed us to the Military Academy at Remberto."

Oflag 64 to Remberto had been quite a journey for H and Clawson and other POWs. Historians acknowledge the difficulty posed for these POWs, "The 1,471 officers and enlisted men under the command of Oberst Fritz Schneider marched to Exin, Poland (24 kms.) where they were supposed to entrain for a new camp in Brandenburg. However, upon their arrival it was discovered that no arrangements had been made for the journey and the group continued on foot to an estate just outside of Exin where they were quartered for the night in cow barns. The weather was below freezing, and the ground was covered with snow. The day's march had been grueling and 186 men decided to hide in the hay lofts and make their way back to the Russian lines. All of these were successful in their escape, and were able to reach Rembertow."[245]

JANUARY 1945, POLAND

H and Clawson entered the large farm house cautiously, feeling famished and frozen, and immediately split up.

H looked around the main level, then made his way back to the silent great room in the center of the house. He took off his hat and unwrapped some of the torn and dirty cloth from around his fingers, stuffing them carefully into a pocket. He knelt down and struck a match and nursed a small fire to life in a massive marble fireplace still stocked with wood.

Filthy and forlorn, H stood up and stretched. He was starting to relax a bit. The snow was falling heavily outside, and it was unlikely they'd been followed to this lonesome place. As he looked around, H was awestruck by the display of wealth surrounding him. Just a poor boy from the mountains of Wyoming, he stood with his jaw dropped.

The grandfather clock stood five feet tall. H could barely make out the time, forever now 5:05, through the shattered glass front. The stair case was elegant with sturdy, carved mahogany railings leading up to the second floor. Ornately framed pictures hung from papered walls. The crystal chandelier cast no light on the long, oak table in the dining room. It was laid as if supper still wafted aromatically from some well-staffed kitchen. China bowls and silver spoons, knives and forks sat on lace and linen. The table seemed ready for servants to scoop and pour and serve rich, creamy cuisine onto plates, and even into the mouths of their well-dressed superiors, who had been Jewish.

They were all gone on this 23rd day of January, 1945. Murdered. Only the rats enjoyed the table now, and they'd pretty well picked it clean.

Harold stared at the scene, dismayed. Here was so much wealth, so much opulence, and no one was left to enjoy it. He'd never seen anything like it—the many rooms, the endless staircase of polished wood, the large glass windows for no one to look out of except himself and Clawson this haunted evening.

They ate canned sardines and partially cooked beans after scouring the house and finding nothing better. They bedded down in front of the fireplace for long-needed sleep, wrapped in blankets Clawson found upstairs.

In the morning, they ate more sardines, stuffing the remaining tins into their pockets and chewed on the remaining beans, now overcooked, that they'd left near the fire.

H and Clawson put on coats, gloves and scarves they'd found in the farmhouse and prepared to face the snowy-cold outside.

A child's portrait hung on the dining room wall. The boy looks down and to the right in silhouette, away from the onlooker. It is an oil painting in brown and

```
     February 1, 1945    RETURNED TO SENDER by Direction
of the War Department, Undeliverable as Addressed.
   2nd Lt. Harold J. Carroll
   Prisoner of War No. 83959
   Oflag 64
   Germany
```

Harold Darling,

I received your letters this morning listing food items you wanted. I have one packed that I'm on my way to mail now and shall fill another this morn and send on with it. I'm glad you said what you wanted.

Pat is just swell and getting cuter every day. Trying to sit up alone now and says hi every time you look at her. She's sure full of old mick—amazing isn't it?

I'm still sewing dresses for Pat—washing. cooking etc. Haven't time for much else. But she keeps me well occupied.

Where shall we go to be alone when you get home? Have some nice place in mind and let's go and fill you up and get you a nice rest. We've lots of time to make up darling so let's do it right.

Gotta dash and take care of some business for you now. Will write again soon.

Patty says, "Tell my Daddy hello and give him my love."

Take care Darling.

Marj

beige with subtle greens and rose. It is not particularly noteworthy, but the portrait of the boy is haunting. His expression is somber, even solemn. The chubby cheeks and long-lashed, demurely downcast eyes emphasize innocence. Was the child a member of the family? Was the picture many years old and the child pictured now long since dead? Could the brushstrokes have been made by a famous hand?

For one raised in austere circumstances, Harold naively believed wealthy people should be able to afford to live. And yet the occupants of this farmhouse had been murdered and their heirlooms abandoned to the Nazis, their food abandoned to the rats, and their oil painting abandoned to a second generation Irish-American escaped paratrooper. H removed the canvas from its frame, rolled it into a baton and secured it inside his jacket. Others collected weapons, and pilfered jewelry. Perhaps H had selected as souvenir something even more tragic.

A few days before D-day. H Carroll, left, Stanley Weinburg, right.

```
   February 1, 1945   RETURNED TO SENDER
by Direction of      the War Department,
Undeliverable as Addressed.
```

   2nd Lt. Harold J. Carroll

   Prisoner of War No. 83959

   Oflag 64

   Germany

Hello Darling,

Just got Patty tucked back into bed so shall take this time to write while she is napping. She's been up for a couple of hours playing in her new play-pen. Also, I got her a darling new high chair since I've gotten home. It has a red upholstered seat and folds into a table and chair. She has a lot of things and all nice but they will undoubtedly see a lot of use in their day.

Received a November 26 letter from you and a September 24 this week. I do hope that long before this you have heard from me and know about Pat. Don't, please don't be disappointed that she isn't a boy for as soon as you see her you will lose your heart over her! I didn't know any two people could look and act so much alike as you and your daughter. It's wonderful! No one has ever said she looks like me but rave about the resemblance to your picture. I've almost finished a dress I started for Pat yesterday. Yes, darling, I'm even learning to sew plus making apple pie. Made two yesterday and was purty good. Yup. Might make a good house wife yet with all of the practice I'm getting. Sure wish it were you I'm cooking for instead of six females. I miss you so much Darling But I'm patient.

Love,
Marj

Photo from canister developed in 2006. Photographer: Harold Carroll

CHAPTER 40

REMBERTOW

There was a large field, without paths and without any type of order. On occasion I found myself squatting at my bodily functions to have a person of the opposite sex squat next to me—and usually bum a cigarette.
~ H Carroll

A DISPLACED PERSON'S CAMP IS A PLACE WHERE THE DESPERATE ARE TOSSED TOGETHER DURING TIMES OF WAR and upheaval. Civilians and military personnel, mainly POWs, shared space in the former Nazi concentration camp at Warszawa-Rembertwo where 200 Jews were shot dead in 1943.

H felt a level of disgust he'd never experienced before upon entering the camp. That disgust grew the longer he was there. He described his experience graphically:

We'd traveled from Brolystok to Grodno to Remberto to Lublin-Vladimir. The Remberto DP [Displaced Persons] camp was Poland's "Hell Hole". When we arrived there we did discover that there was something of a chain of command. As a matter of fact when Clawson and I arrived we were greeted by someone (American Field Grade) who informed us that we had been exposed to a contagious disease in one of the towns we had passed through and therefore would have to be "isolated."

We were then led way back into the bowels of the huge building that housed, God only knows how many thousands of Displaced Persons of every nationality on earth. We were shown into a "cubicle" in which were 30 or 40 unidentifiable people, who were defecating and copulating as if there were no tomorrow. We refused the invitation to become part of that jolly group and informed our American benefactor that we were either leaving the Compound or would accept acceptable quarters—for a while. I think there was partial grouping of Americans in the DP Compound; Clawson and I shared a fair size room with about 5 or 6 other American Officers.

In passing, it might be well to mention the 'Sanitary Facilities' at the DP Compound. There was a large field, without paths and without any type of order. On occasion I found myself squatting at my bodily functions to have a person of the opposite sex squat next to me—and usually bum a cigarette. The one shower that I remember—also de-lousing—was entirely co-ed.

To put it mildly, our stay in Remberto was 'no picnic.' This old Military Academy had been made into a D.P. Camp and there were people of every nationality there—and not just military. Clawson and I tried to make our way out of there—unsuccessfully—and would have gone to Moscow or Mermansk or anyplace but the Russians wanted nothing to do with us.

We did attempt to make our way to the Black Sea by taking certain routes out of Warsaw-Remberto but each time we were stopped at some Check-point by Russians and turned back. Not arrested, not detained and most certainly not entertained—just informed Americanski in Remberto. I feel quite certain that both Col. Drury and Col. Millett were at the DP Compound but cannot remember what contact I had with either. We did leave Remberto separately after hearing many of the rumors about the arrival of the Embassy from Moscow and their plan to provide transportation.

We were there in that D.P. Camp until Feb. 21st; then started walking and ultimately caught up to a Freight Train on its way to Odessa. [There was a rail-head at Lublin and there they joined a train load of Allied Ex-prisoners.]

On Feb. 22nd, we boarded the now familiar 40 x 8's.[246]

Photo from canister developed in 2006

CHAPTER 41

ESCAPE FROM ODESSA

There was absolutely nothing to identify us as Americans. We were wearing raunchy fur caps, Polish great coats—or worse—and underneath it all the same uniform that we had left England wearing on Jun 5th. But the American crew recognized us by our walk or mannerism or instinct. ~ H Carroll

WHEN CLARENCE MELTESEN INTERVIEWED THE EX-POWs WHO'D PASSED THROUGH ODESSA HE RECEIVED A SPLIT vote regarding conditions there, "a few say it wasn't so bad, more say it was pretty restrictive, and one or two who came by way of Kiev area seem to have had a very hard time."[247]

H and Clawson arrived in Odessa on February 28th and found the environment of the Russian run compound oppressive. H was particularly insulted by Odessa's class structure which offended his Scots-Irish sensibilities.

"Odessa was very large, segregated building that, I believe, had previously housed a Military Academy. By segregated I mean that the Enlisted men were separated from the Officers. My recollection tells me that it was a very gloomy complex of buildings surrounded by a tall stone or rock fence. Most of us saw nothing nor heard anything of the Representatives from the Embassy at Moscow but were told that if we were to put our name on a list together with our rank that we would be scheduled, by the Embassy, for shipment 'out.'

"The food was terrible—'Kasha which was a combination of boiled rice and cabbage; and, for a treat augmented by a 'Fish Head'. I well remember the icy stare of that damned fish in the center of that plate of yuk. The compound was obviously well guarded by the Russians and variation from our activity of sitting and waiting was not allowed.

"It should also be noted that we were essentially 'prisoners' while in Odessa. Clawson and I escaped…

"Clawson and I soon determined that sight-seeing was preferable, located an unguarded door in the high stone wall that surrounded the compound, left the compound and made our way to the docks. Excuse a diversion here…it was somewhat noteworthy that at that time the first ships were in Odessa harbor since 1938—I think—These were two American manned Liberty Ships that were unloading heavy equipment. Rather the Russians were unloading it while the Americans stood at the rail a great deal of the time enjoying the antics and ineptness of the Russians. As we walked down the docks toward the one ship several of the crew were standing at the rail.

"There was absolutely nothing to identify us as Americans. We were wearing raunchy fur caps, Polish great coats—or worse and underneath it all the same uniform that we had left England wearing on Jun 5th. But the American crew recognized us by our walk or mannerism or instinct—distracted the attention of the Russians guarding the gangplanks and managed to get us aboard.

"We spent the day and part of the night with them. They saw to it that we had a shower, gave us clean underwear, fed us a banquet complete with a 'special' occasion bottle of Old Taylor. The Captain offered us a passage aboard back to the States but advised that their schedule included traveling through the Mediterranean, through the Azores, into the British Isles and that it would likely take as long as nine months for them to reach the States. We reluctantly declined and made our way back to the compound.

"The next day we again went into Odessa near the docks. Here we ran into a U.S. Navy CPO [Chief Petty Officer] who was stationed there. That was CPO Walleck. He loaded us into his Jeep and took us to his apartment where we enjoyed a dinner of, by gosh, Pork Chops and all the dressing. We probably made other trips out of the compound—I do not recall—I don't know the name of the Ship.

"Clawson and I prowled around Odessa a lot…We did ultimately decide that the fastest and most logical way to leave Odessa was by way of that Ship provided by the American Embassy."

Transportation home was eventually arranged by the American Embassy in Moscow.[248]

Mar 23, 1945

AIR LETTER

MRS H.J. CARROLL

THE FOLLOWING CERTIFICATE MUST BE SIGNED BY THE WRITER: —

I certify on my honour that the contents of this envelope refer to nothing but private and family matters.

Signature   Harold J Carroll

2nd LT INF

O-1308095

Dear Marjorie,

Am free and in Allied hands, safe and in good health. Within the near future a more detailed letter will follow. Trust that you are well, may see you soon.

Harold

March 1945

WESTERN UNION

MRS MARJORIE H CARROLL=

AM PLEASED TO INFORM YOU RREPORT RECEIVED FROM UNITED STATES MILITARY MISSION IN MOSCOW STATES YOUR HUSBAND SECONG LIEUTENANT HAROLD J CARROLL PREVIOUSLY REPORTED PRISONER OF WAR HAS BEEN RELEASED FROM A GERMAN PRIISONER OF WAR CAMP THE WAR DEPARTMENT INVITES SUBMISSION OF A MESSAGE NOT TO EXCEED TWENTY FIVE WORDS FOR ATTEMPTED DELIVERY TO HIM MESSAGE SHOULD BE ADDRESSED TO CASUALTY BRANCH AGO ROOM TWENTO FIVE FIFTEEN MUNITIONS BUILDING FURTHER INFORMAITON WILL BE FURNISHED WHEN RECEIVED=

J A ULIO THE ADJUTANT GENERAL

April 1945, 1945

WESTERN UNION

MRS MARJORIE H CARROLLL=

THE CHIEF OF STAFF DIRECTS ME TO INFORM YOU YOUR HUSBAND HAROLD J CARROL IS BEING RETURNED TO THE UNITED STATES WITHIN THE NEAR FUTURE AND WILL BE GIVEN AN OPPORTUNITY TO COMMUNICATE WITH YOU UPON ARRIVAL=

J A ULIO   THE ADJUTANT GENERAL

WHEN I GROW WEARY

Robert E. Carroll

When I grow weary in the fight
To win my daily bread,
When roads ahead are shrouded in
A maze of fear and dread,
I close my eyes and drift away
In thoughts to distant lands,
Where nature's wonders still remain
Unsoiled by human hands,
To breathe again the mountain air
And walk the rugged trails
That wind among the lofty pines
To cool secluded vales,
Where I have often paused to watch
The slowly drifting clouds
Enfold the stately mountain peaks
With lustrous fleecy shrouds,
No magic tones of flute or harp
Nor noble works of art,
Can still the raging storms within
Like moments spent apart
Far out beyond the traffic lanes
In some quiet forest glade,
And feast my soul in raptured awe
On things that God has made.

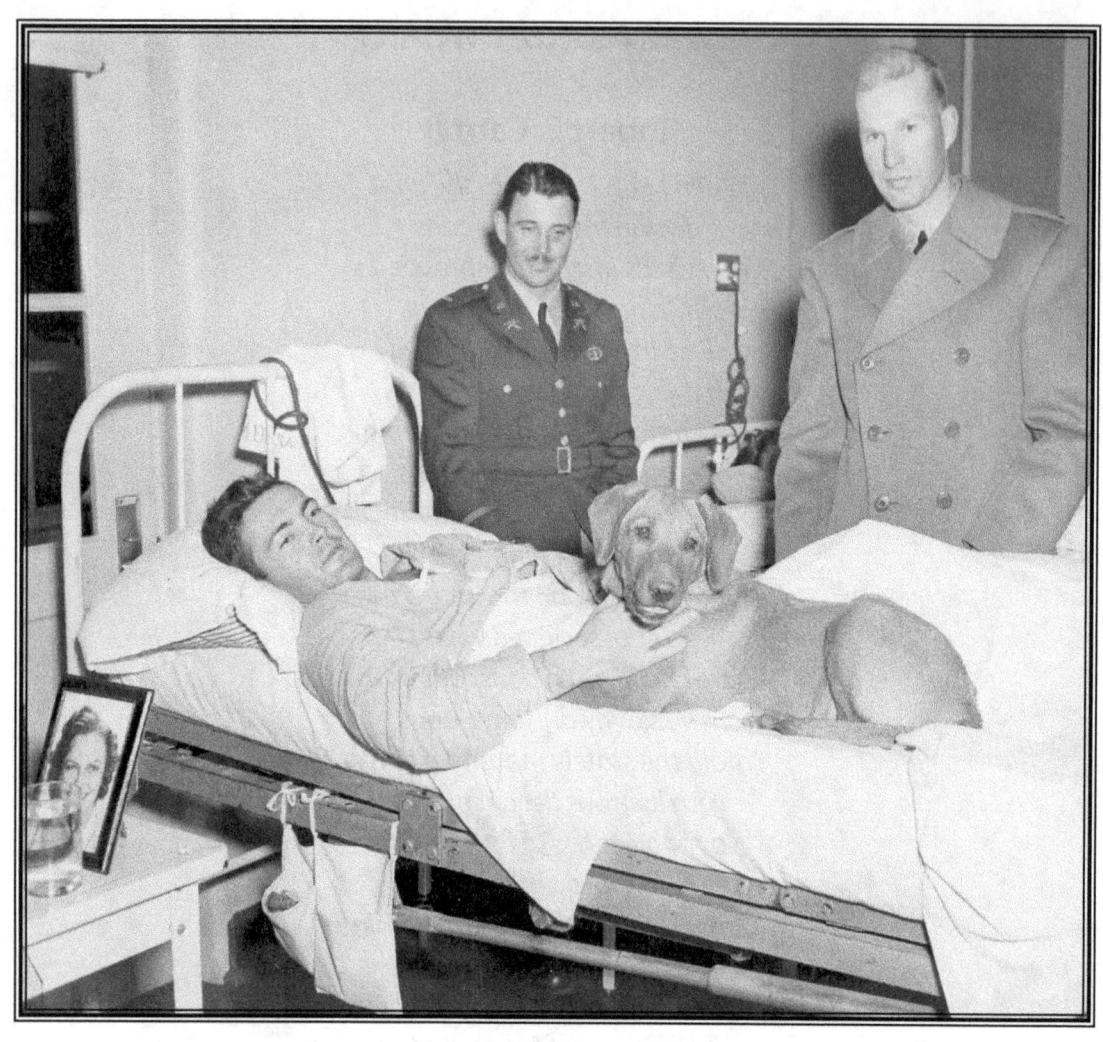

Bob Carroll, right, with injured soldiier in England before the invasion.

CHAPTER 42

HOMECOMING
Am free and in Allied hands, safe and in good health. ~ H

Boys going to war look completely different than returning veterans. When H enlisted in 1941, his picture taken just before leaving home shows him dressed in khaki, teeth blazing white, eyes sparkling, confident and youthful.

"When we arrived at the Oflag on Sept. 18, I weighed 106 lbs. I had gone into the invasion at 160." Harold's appearance had drastically changed. He "resembled only a caricature of his former self. An identification picture of the lieutenant in the camp reveals that fact clearly."[249]

H had experienced an eleven-month ordeal that began June 6, 1944 and did not end until April, 1945. Between July, 1944 and April, 1945, H spent time as a POW at camps Chalon Sur Marne, France; Stalag III B, Chalon, Germany; Stalag XII A, Limberg, Germany; Oflag 64, Schubin Poland; and Odessa, Poland. H faced a firing squad at least three times. He and other POWs were endangered by attacks from their own allied planes and spent hours cramped together into overcrowded boxcars. H marched hundreds of miles on foot in worn-out boots, wounded, cold and hungry. He'd been interrogated, beaten and shot. And yet he knew he was lucky—he had seen the horrors at Praga, a Nazi death camp, and he was alive to tell of it.

H was finally going home. He'd had a belly-full. "Clawson and I left Odessa on March 2, 1945. I first reached an American Military Installation in Port Said, Egypt on about March 10, on board the ship Camp Myles Standish. We arrived in Naples on March 20th and I was hospitalized for several days."

Clawson managed to get a ride back to France to re-join the 82nd Airborne—and, "When I was deemed fit to travel, I joined a group and left Naples on March 29."

In a letter from Carl Clawson to Deryk Wills, Carl recounted that Harold had "a very bad arm and was sent home from Naples—I rejoined B Company in Germany and crossed the Elbe only to be wounded a day before the war ended and was sent home from the hospital."

The Drop, photographer unknown, from loose photo card circa 1942

SPRING 1945

New Orleans was one big party, and it wasn't even Mardi Gras. Bands were playing, and people were singing. They lined the streets of the French Quarter and the docks. Bonfires burned and people danced late into the night. Art was discharged into the happy celebration at the harbor and Mildred ran through the crowd to greet him.

She'd left their young son, John, with her parents in Oklahoma for this very special reunion. "It was wonderful to be dancing to big band music, eating at great places and seeing beautiful buildings, but when those discharge papers were signed, Art wanted to see his son!"

Mildred had continued her work in the Platt Studio in New York until she was seven months pregnant. "Mr. Platt thoughtfully offered upstairs rooms with a nanny—he thought it would be the perfect place for little Butch or Butchess. But I thought that Ada, Oklahoma would be better. John Arthur Carroll was born in Oklahoma May 18, 1945."

But the couple didn't remain in Ada. After a short visit Art and Mildred moved their household up to Sheridan, Wyoming. "It was a great reunion!" she remembered, "with Harold, Marj and Patricia Lynn," (born September 29, 1944). Art got a job at the hardware store and he and Mildred rented a little house from Col. Lamb and his wife who were still overseas. It was too late to get into classes at the University of Wyoming, so Art made plans to enter the university the following June, joining the summer session. His dream of an education was still alive and well.

One evening Milly and Art sat down at a Sheridan dinner and dancing club with some friends. Mildred had to stifle a groan when she saw the band coming. She was raised a small town girl, but she'd spent the last five and a half years in NYC, for goodness sake!

She covered her gaping mouth with her hand and looked away. "Here came the orchestra. They were all in cowboy shirts and cowboy hats and cowboy boots. And I thought, oh, my gosh, it's gonna be a western band." Then they sat down and picked up their horns and played the sweetest Glen Miller, Artie Shaw, everything we liked!" Art was such a good dancer that Milly would have danced with him no matter what they were playing. "I'd already told him I married him because he was a good dancer. All the girls wanted to be on his arm."

The couples did a lot of socializing, especially bridge playing, after the boys returned from service. Mildred and Art and Harold and Marj gathered around the Bridge table regularly. The children were usually well-behaved but occasionally there were tears or tantrums and a child would throw themselves upon the floor. Then someone would announce loudly to the others: "Leave her lay where Jesus flung her!" and everyone would laugh.

Mildred remembered one particular bridge game shortly after arriving in Sheridan. They were playing cards in front of the large bay window overlooking the street. A loud whooshing sound was heard just before something smashed through the glass. Art and Harold both acted instinctively. They shoved their wives unceremoniously under the table by placing their hands on the girl's heads, and pushing them down.

No grenade. Teenagers had thrown a dead rabbit through the window. Art and Harold later apologized for their reflexes.

H had encountered similar problems with his reflexes before. When H got back to the states, prior to actually being sent home, but after being released from the hospital, he went to an officer's club one night. He was standing, sipping his drink on the second level of the bar when an airman came up to him, introduced himself and started talking. He asked what outfit H had been with and where he'd been.

"I'm in the 82nd." H wasn't feeling up for a long-winded talk, "Paratroops."

When told by H he had been part of the invasion force, the airman said with a grin, "Well, hey, maybe I was the one who dropped you!"

H replied, "That so? You like to fly?"

"Sure thing!" said the pilot.

H then picked the guy up and as he threw him over the balcony said, "then fly boy!" H was having a bit of trouble returning to the good ol' Harold everyone back home remembered.

Across the Atlantic, Hitler suffered a little PTSD (post traumatic stress disorder) of his own. His hair had gone completely gray, his back was stooped, and he had difficulty walking. His eyesight was poor, and overall he looked much older than his years. A nervous disorder he'd suffered since WWI left him with a tremor in his left arm and leg.

Many were shocked by his appearance. One man remarked: "It was a ghastly physical image he presented. The upper part of his body was bowed and he dragged his feet as he made his way slowly and laboriously through the bunker from his living room... If anyone happened to stop him during this short walk (some fifty or sixty yards), he was forced either to sit down on one of the seats placed along the walls for the purpose, or to catch hold of the person he was speaking to... Often saliva would dribble from the corners of his mouth... presenting a hideous and pitiful spectacle."[250]

MY CREED
By Edgar Guest

To live as gently as I can
To be, no matter where, a man
To take what comes of good or ill
And cling to faith and honor still
To do my best, and let that stand
The record of my brain and hand
And then, should failure come to me
Still work and hope for victory.

To have no secret place wherein
I stoop unseen to shame or sin
To be the same when I'm alone
As when my every deed is known
To live undaunted, unafraid
Of any step that I have made
To be without pretense or sham
Exactly what men think I am.

To leave some simple mark behind
To keep my having lived in mind
If enmity to aught I show
To be an honest, generous foe
To play my little part, nor whine
That greater honors are not mine.
This, I believe, is all I need
For my philosophy and creed.[*]

---

[*] "Over the years I have worked with quite a few people, but only one Harold Carroll—a special guy. He was always there with a helping hand, a ready smile or a listening ear. He is 'one of the best'. The following poem seems symbolic of you, Harold." Evelyn Houser.

Smooth Landing, photographer unknown,
from loose photo card circa 1942

HEADQUARTERS 505th PARACHUTE INFANTRY
Office of the Regimental Commander
A.F.C. 469, U.S. Army
15 October 1945
To: Lt. H.J. Carroll, 0-1308085
45 Swan Street
Sheridan, Wyoming

Dear Carroll:

I've been trying to find out exactly what your status is, but so far have no information except that you were shipped to the zone of the Interior. I'm going to send to you one of the two letters you left with Mr. Lemkowitz.

I took the liberty of opening the one which you addressed to yourself and the contents are enclosed herein. The other marked "Mail in event of my death" I have personally destroyed since you are very much alive. I hope I have done that which you desired.

Things have changed immeasurably since you left and the Division is preparing to move to the States soon. We should be home well before Christmas and I hope our paths will cross sometime in the future.

For your information the Division has been awarded the Belgium Fourragere 1940 and the "Militaire Willems Order" of the Netherlands for action in the Bulge and Holland respectively. We're hoping to get the French Fourragere soon—if we do I'll let you know.

Must close now and get some of the papers off my desk. Will give you more information the next time I write.

Good luck and thanks for turning out a superior brand of soldiering. Men like you made the Regiment what it is and has given it a record second to none.

Sincerely,

William E. Ekman
Colonel, Infantry Commanding[253]

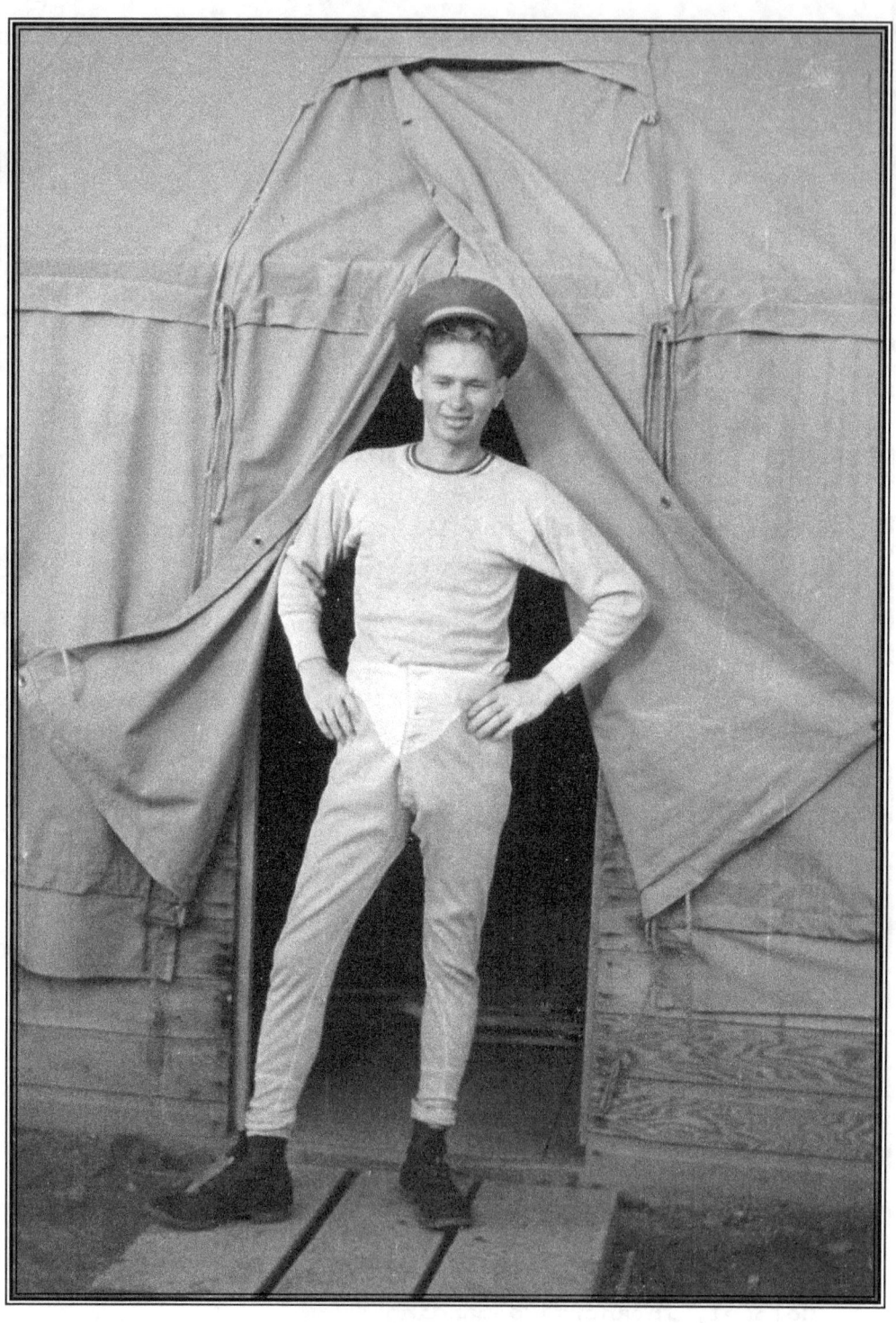

Photo from canister developed in 2006. Photographer: Harold Carroll

CHAPTER 43

ZIPPOS AND HOSIERY

The word "Zippo" was created by Mr. Blaisdell in 1932. He liked the sound of the word "zipper" so he formed different variations on the word and settled on "Zippo," deciding that it had a "modern" sound. ~ Zippo History online.

Harold arrived back in the good ole U.S. of A. in Boston Harbor on April 15, 1945. "You've heard of soldiers kissing the earth when they step ashore. I didn't do that but I certainly felt like it," he declared. "Roosevelt died April 12, 1945, while we were aboard ship and Truman became President."

Things changed after victory was announced. Adolf Hitler committed suicide. Marjorie was finally able to replace her worn hosiery and Patty was able to ride her new tricycle as rubber for tires became abundant. After arriving home, Harold transposed the etchings from his Zippo lighter. "I do have many dates and the towns that we were in on our travels through Poland—and before we were finally taken to Oflag 64. I had first scratched down names of towns and dates on a Zippo lighter I had been carrying for about 10 or so years. Later I transferred those dates and places to scraps of paper."

Harold tried to retrace his steps and review the route he'd traveled, outlining possible paths on copies of the many maps he collected. "We covered all of the territory between Cherbourg and Odessa. Our journey across Poland was somewhat confusing. First of all Germans had renamed many of the towns and some of the names in my notes are slightly vague. However, I have tried to compare them with later maps of Poland

and feel that I have pretty much mapped out our route."

Harold carefully preserved his notes. He read the book *Sage* with a great deal of interest. "Comparing names of towns and dates, etc., I find that Jerry Sage [the author] traveled much the same route that Clawson and I did, however, it does appear that there is just a bit of fiction in Sage and, if his listing of towns is anywhere near accurate he sure as hell rambled around a lot. He escaped nearly the same place we did—could have caught a ride on the same train we did but then he went way back up north to Torun quite a distance, and then back south to Brzesc Kwawki—it doesn't make sense. (I find that his route and ours was similar, except—it seems that he did wander all over for a while, having doubled back from Wrzesnia clear back north to Torun.) Also, the 'Sage' account of his visit to Port Said somewhat coincides with ours."[251]

Harold came home with many mementos of his time in service. From Oflag 64 he had the Program from the play he attended: "The Man Who Came to Dinner." He had a menu from Christmas dinner 1944 printed by Druck. He had his Kriegy I.D. card, and letters and cables to and from home. He also came home with a copy of "The Yankee Kriegies" by Colonel C. Ross Greening and Sergeant Angelo M. Spinelli and the Red Cross Publication "Prisoners of War Bulletin" for March, April and May 1945. And for some reason he hung onto the Paratrooper Training Guide.

Another memento Harold returned home with was severe post traumatic stress disorder or PTSD; he had nightmares for many years after the war. And he returned home without an older brother. Three boys had gone to war. Only two returned.

THE ALL AMERICAN NEWS
THIS SPECIAL EDITION IS COMPOSED OF ITEMS PREVIOUSLY PUBLISHED IN THIS PAPER
WELCOME HOME AIRBORNE 82nd
82nd DIVISION ASSOCIATION
26 EAST 39th STREET, NEW YORK 16, N.Y.
SPECIAL (January 12th, 1946) EDITION

D DAY AND PLUS "Some new names have been added to the battle-scarred flag of the 82nd. "Gela," "Salerno," "Isigny" and "Ste Marie Eglise." The Airborne has done all and more than we expected of them. The 101st Airborne, or Eagle Division took some part in the latest operation and in this division we, too, have a proprietary interest. For, in the 101st, you will find a 327th Inf. and a 321st Field Artillery. You have read so much of what these Airborne Divisions have done in the present invasion that anything said here would be a repetition. We will simply dwell on the strategic phase of the invasion. The great port of Cherbourg with its docks and rail facilities was the immediate objective. It sits on the northern tip of a peninsula, or neck of land, approximately twenty miles wide and thirty miles from the tip to the torso. The invasion was probably the most ambitious single effort of the war, encompassing a naval, aerial and amphibious assault. The initial breach was roughly between Isigny on the east and Caen on the west. Six hours before the arrival of the amphibious force, the 82nd and the 101st were dropped from the skies around Isigny, and the British 6th Airborne was dropped around Caen. They immediately covered bridges, canals and roads so that Nazi reinforcements could not be sent to the beachhead. The Seaborne troops, fighting through a staggering frontal attack after forty-eight hours made contacts with the Airborne troops on each flank. Thus the infantry landing forces had made to order for them a solid protective right and left flank even before they had landed. The West, or British flank, became the anchor flank, resisting all German pressure to break through, but doing purely defensive fighting. The east, or American flank, became one of movement to knife through the Cherbourg Peninsula at its narrowest point. In this, the 82nd was in the forefront most of the way. But, when the movement veered north toward Cherbourg, the 9th division took over the spearhead of the offensive from the 82nd, which has probably been sent into reserve for future use. At this writing the first phase of this action, the capture of Cherbourg, is accomplished. This will be used as a debarkation area for invasion troops. The west flank held by the British and Canadians at Caen will become active, having one, or maybe two objectives. The first or more important to shut off attack from the rear and eliminate all the Robot invasion coast stretching from Cherbourg to Calais if it cannot be successfully blasted out of existence by air power. The second would be a drive towards Paris and the interior and an eventual contact with the 5th and 8th Armies in Italy. The line of encirclement will continue to grow with each operation until the bands of steel strangle and choke Nazism right in its home lair."

SOMETIME AFTER THE WAR

Patty Carroll holds her daddy's hand as they wait for a break in the traffic so they can cross the busy street. She's excited by the city and the experience of spending time alone with her dad. Denver seems like a long way from home, but he holds her hand firmly, and she feels safe.

They cross finally, approaching a low brick building with a brightly lit sign. Daddy holds the door open for her and they enter the large, dimly lit room dominated by a billiards table and a long, shiny wooden bar lined with green vinyl covered stools. She holds her daddy's hand tighter as they approach the bar.

A short, muscular man stands behind the bar by a sink of running water. He is wiping a beer glass with a white cotton towel and taking care to buff the glass spot free. He looks up as Harold and Patty approach.

His mouth drops open. The glass falls from his hands smashing on the floor with a tinkling crash.

"You're dead!" he blusters, red faced. Patty slips behind the protective form of her father as the man falls loudly to the floor, unconscious.

Later, over stiff drinks, the bartender insists, "I can't believe my eyes! You saved my life! Then I saw you killed, Sarge!" He pours another Coca Cola for Patty. She sits tall on the barstool, taking in the surroundings, and sips the cold, sweet drink from a straw. "I remember you pulled me up yelling, 'C'mon, buddy, we gotta get outta here.'"

The pair clinked glasses over shared memories while Patty looked on.*

---

* "I vaguely remember waking to him hollering in the night and mom picking out shrapnel which would painfully migrate to the surface as angry infected boils, well into my pre-teen years, a good 10 years later," remembers Pat Carroll Cleveland.

Bob Carroll

CHAPTER 44

MISSING IN ACTION
*There be of them, that have left a name behind them,
that their praises might be reported. And some there be,
which have no memorial. ~ Britain's Homage To 28,000 American Dead,
dedicated to the memory of Lt. Col. R. C. Carroll, U.S. Army.*[252]

IN THE FOG OF WAR, RECORDS WERE LOST AND CONFUSION ABOUNDED. SOME THINGS WERE UNDERSTOOD AND PIECED together years later with the clarity of hindsight, or after secret papers were revealed. Some things still remain a mystery.

What happened to Bob? In his book *D-Day with the Screaming Eagles* George Koskimaki looked for answers to this question and others: "What had happened to the men of the 1st Battalion who had the original assignment of capturing the locks? Why were so few of them available for duty on the first day at La Barquette?... The battalions were scattered over the countryside. It would take many days to get the survivors back to their units. By D-plus 2, less than half of the 2300-man force was to assemble. For the 1st Battalion, the situation was critical. The battalion had to be reorganized. Its command personnel were gone... ambushed shortly after getting together near a crossroad and all three lost their lives."[254]

1st Battalion Commander of the 501st PIR, 101st Airborne, Lt. Col. Robert C. Carroll, was killed and his executive and good friend, Major Philip S Gage, was wounded and captured on D-Day. His entire headquarters staff was missing as were the commanders of Able, Baker, Charley, and Headquarters Companies.[255]

The 101st Airborne website reports, "the battalion commander, Lt. Col. Robert C. Carroll, and two of his staff officers, Captain Thomas Chastant and Lt. John W. Atkinson, Jr., had run into a German ambush at a crossroads before daylight. PFC Leo Runge remembered, 'I was near Col. Carroll when he was killed before daybreak.' (The bodies of the three officers were picked up by Lt. Sumpter Blackmon in a burial detail a few days later.)"

Harold did not believe the reports of Bob's death. He retained hope that Bob survived, that they might be someday reunited. He searched for his big brother whom he loved and admired, and resembled so closely. There were reasons Harold believed Bob still lived. He remembered the German soldier, old hawk nose, who approached him while marching enroute to that first prisoner of war camp who told H he'd seen his brother —must be his brother they look exactly alike—marching toward Chalon Sur Marne, France. That wasn't an illusion. And on another occasion after the invasion someone had mistaken him for Bob. It couldn't be true that Bob was killed as reported on D-Day.

Hell, Harold had been reported "Killed in Action." People were seemingly resurrected who'd been "Killed in Action" before. Clearly, Bob had not returned home after the war, but that was because he was captured, H imagined. Bob had likely been trapped behind the enemy line, and then trapped behind the Iron Curtain at the mercy of the Soviets, like so many others.

According to an article in the Wall Street Journal on August 13, 1987, the Soviets kidnapped 16,000 American servicemen at the Russian port of Odessa in 1945 and they "lost" another 3,000 American fighting men in Poland at about the same time. The 16,000 Americans the Russians are said to have kidnapped were those they had previously liberated from German prisoner of war camps.[256] It was newspaper articles like this that kindled Harold's hopes and stirred his anger.

Adding to the fog of war that prevented families from discovering the the fate or whereabouts of missing soldiers, according to Bill Paul, was the deliberate slavery of former Americans, and others, by the USSR.

Harold hung on Paul's words describing a top secret report written by U.S. Maj. Gen. John R. Deane, who was attached to the U.S. military mission in Moscow. the topic was the Soviet treatment of Americans liberated from German POW camps. Paul said that "On March 12, 1945, W. Averell Harriman, U.S. ambassador to the Soviet Union, sent a top-secret message to President Roosevelt." The message informed the President that the Russians were holding American servicemen illegally.

"John Noble met Secretary Schultz (then Secretary of State George Schultz) in Washington in 1984, Noble told him that Americans are being held in Soviet camps, to which the secretary replied, 'I know that they're there.'"

"Noble should know. The Germans imprisoned him in 1939 and when the Russians liberated him in 1945 they took him to Russia, where they held him until they released him in 1955."[257] Noble had been taken prisoner after the Invasion of France and then kept prisoner behind the Iron Curtain for ten years. If Noble could still be alive after all of that, then so could Bob, Harold imagined.

"The problem of accounting for POW/MIAs was complicated by the fact that the Soviets were just as uncooperative in the repatriation of the millions of displaced civilians. In Europe, as well as in the Far East, the Soviets guarded a sea of prisoners—human capital and slave labor in their view—who were not only Allied and Axis POWs, but also hundreds of thousands of displaced Western European citizens."[258]

Could Harold have been right? Could Bob have been captured instead of killed by the Nazis and then enslaved by the Soviets and made to work in the mines? We will never know—just as we will never know what happened to many others who served, or even who they were.

In Arlington, Virginia, the Tomb of the Unknowns attracts many tourists. Words inscribed in marble state: "Here Rests in Honored Glory an American Soldier Known But to God." Inside lie the remains of unidentified American soldiers from WWI, WWII, the Korean War and Vietnam.

The cemetery at Saint Mere Eglese, France is filled with row upon row upon row of white crosses placed in memorial of soldiers fallen on French soil during WWII. One of the crosses, in the sea of many, carries the name Robert Collin Carroll, 1st Battalion Commander of the 501st Parachute Infantry Regiment of the 101st Airborne.

SPRING 2014

Bob's story is similar to the tale he loved to tell about Davy Crockett years earlier. Bob and five or six of the bravest and finest American paratroopers were all captured during the invasion of Normandy. They were dragged in front of Hitler's top SS general. The general was angry, still bleeding from wounds caused by Bob and his paratroopers, and he couldn't believe his Axis soldiers had brought these Allied soldiers into his presence alive. He ordered them all bayoneted and then shot.

And yet some still claim that Bob Carroll survived the Invasion. He was taken to the salt mines deep in the Soviet interior, and made to work as a slave. He may survive there to this day. He never would have regretted his actions. He had lived true to Crockett's words: "Be always sure you are right, then go ahead."

But by gum, he'd be old today.

B-company, a few days before D-day, photographer: Harold Carroll

AFTERWORD
Who knows? Who hopes? Who troubles? Let it pass! ~ Wilfred Owen

To record the WWII experiences of Bob, Art and Harold accurately, I needed to do more than list a series of events—birth, marriage, battles, death—but to also paint a picture and convey an understanding of the elements that created their characters and their particular response to their time and place in history. This task required looking back at their family origins and attempting to set judgment aside, and see beyond the stale facts to a place of conjecture—myth if you will.

In juxtaposing history and myth I hoped to get at the bigger picture, the picture that facts alone can't tell.

"Myths are lies, though lies breathed through silver," postulates C.S. Lewis as he walks along a river discussing the nature and purpose of myth with J.R.R. Tolkien.

"No," disagrees Tolkien, "They are not. There are truths that can be expressed only through the language of myth."[259]

I don't believe that myth perpetuates lies, instead myth illustrates underlying relationships and presents us with possibilities. Myth is not about describing what did happen as much as what can happen and how it makes us feel.

I was surprised at the theme of resurrection that continually came up (rose in the prose) as this manuscript developed. First, the belief that Davy Crockett survived the Alamo and worked the mines as a prisoner in Mexico; second, the belief that the Australian POWs on the Montevideo Maru survived the ships' demise at sea and worked as slave laborers in

Japan or on Japanese held islands; third, Harold's tenacious belief that his brother Bob had survived the Invasion and was a prisoner of the Soviets, laboring behind the Iron Curtain.*

I'm not sure why grieving family members would prefer their lost loved ones were alive and living as slaves rather than long-dead and pain-free. I guess it boils down to hope and the possibility of eventual reunion.

J.R.R. Tolkien observed that during the 20th century more lives were lost to war than in all the previous centuries combined. He regarded modern warfare as "uniquely evil." C. S. Lewis, in contrast, believed wars would always exist and asserted that "modern combat is more vicious than ancient enmity only because we have more destructive weapons at hand."

World War II took the lives of over 60 million people. It remains the deadliest conflict in human history to date with sixty-one nations and more than one hundred million military personnel involved. WWII involved the complete mobilization of U.S. economic, industrial, and scientific capabilities, and it changed the world in significant ways—ways that can never be retracted, including the development and unleashing of nuclear power.

Some doubt the need for America's involvement in WWII, or any other war. And yet, as Emannuel Allain, owner of a museum in Saint Mere Eglese, France pointed out to me in a letter, if it weren't for Bob, Art and Harold, and other American soldiers and supporters, "I would have grown up speaking German."

I returned home after my father, Harold Carroll's, funeral and a short visit with my sisters. In my possession were two medium-sized boxes my dad had stored in the shed. They were crammed full of notes, clippings, photos, and film that he had collected and saved over his lifetime.

The boxes sat in my closet at home and occasionally I would go through and sort the musty time-forgotten memorabilia into scrapbooks. Those scrapbooks eventually turned into the words you are reading.

This now is for you. Explore it. Add to it. Correct it. It is a canvas on which you are reflected. Do you see yourself in the picture? Remember, always look back seven generations and forward seven generations. You are impacted by the past just as you impact the future.

Peace.

M. Carroll

---

* Of note: there are still reported sightings of camels in the Southwest today, according to Hi Jolly's prospector friend, Bill.

Sheridan, 1930

ASLEEP
By Wilfred Owen

Under his helmet, up against his pack,
After the many days of work and waking,
Sleep took him by the brow and laid him back.
And in the happy no-time of his sleeping,
Death took him by the heart. There was a quaking
Of the aborted life within him leaping...
Then chest and sleepy arms once again fell slack.
And soon the slow, stray blood came creeping
From the intrusive lead, like ants on track.

Whether his deeper sleep lie shaded by the shaking
Of great wings, and the thoughts that hung the stars,
High pillowed on calm pillows of God's making,
Above these clouds, these rains, these sheets of lead,
And these wind's scimitars;
— Or whether yet his thin and sodden head
Confuses more and more with the low mould,
His hair being one with the grey grass
And finished fields of Autumn that are old...
Who knows? Who hopes? Who troubles? Let it pass!
He sleeps. He sleeps less tremulous, less cold
Than he who must awake, and waking, say Alas!

APPENDIX A

The Tale of Abraham's Ghost
(Extrapolated from many versions on Ancestry.com and other websites)

Strange stories passed down through the centuries of ghostly hauntings and buried gold surround Abraham Kuykendall. Certainly he'd rather be remembered for his valor as a fighting captain in the North Carolina Militia during the American Revolution rather than the oft recounted tales of his marriage to a beauteous young wife Bathsheba, his large and popular tavern, and his desire to be paid only in gold or silver coin.

The story is told of an old Abraham, greying and long in the tooth. His first wife, Elizabeth, had died giving birth to their eleventh child. A rich man, he soon charmed the young and very alluring Bathsheba. They wed and lived on his thousand acre settlement in Flat Rock, North Carolina where he built an unusually large and busy tavern around 1800. Bathsheba gave Abraham four healthy sons and she helped entertain guests and spent his money lavishly on fine clothing and jewelry.

The tavern sat prominently off of Old State Road, a busy thoroughfare that teemed with people driving herds of cattle, horses, and mules from Kentucky and Tennessee to the markets in South Carolina and Georgia. Thanks to Abraham's hard work and Bathsheba's attention to detail, the tavern offered the very best of accommodations to weary travelers who came to expect soft beds, great grub and grog—hand cut steaks and whiskey made at Abraham's own still, and secure holding pens for their horses and livestock. For this, many were willing to pay handsomely. Abraham's tavern became well-known and its reputation drew many guests from far and wide.

Abraham required all visitors to pay in gold or silver. Soon he had quite a pile of money and began to fear for thieves. He stored his earnings in strong boxes locked with chains and padlocks in a secure room in the tavern. One night, Abraham got the notion that the contents of his overflowing strong boxes required safer storage. And so, with only a sliver of moon for light and the wind whipping up a ruckus in the trees, old Abraham woke his most trusted slaves, whom he called servants.

"Get up. I want you to bring the wash pots up the hill," he ordered the two baffled young men.

They did as they were directed, carrying the large iron wash pots to the room where Abraham kept his strong boxes. After having them transfer gold & silver from the strongboxes to the wash pots, Abraham blindfolded them both. They carried the wash pots down the road and into the forest following his words and footfalls, Abraham led them carrying a single torch.

They left the main road and walked into the forest, stumbling occasionally and groaning from the weight they carried.

"Put them down here," Abraham spoke confidently and removed their blindfolds. Offering them the spades he'd been carrying, he ordered them to dig a hole under a bent white oak tree near the stream. Sweat poured off of the men as they dug deeper and deeper. Finally, Abraham stopped their digging and they placed the wash pots into the hole and covered it, and hid the spot with leaves and twigs. Once again blindfolded, he led the servants back to the darkened tavern.

"Speak of this at your peril," he warned them both.

Many years later, when Abraham was 104 years old, so the story goes, he entered into a business arrangement for which he needed more money than was available in his strongbox. He took a shovel and left, walking down the long road away from the tavern and then into the thick forest, all alone. Hours and then days went by, but Abraham never returned. A search posse was formed and he was finally found face down in a mountain stream. It was assumed that old Abraham had fallen and hit his head, drowning in the cold mountain stream.

When word got out about his death, the two frightened servants came forward. They told the family about the dark and windy night and the wash pots full of gold and silver coins buried deep beside a large white oak near a mountain stream. Many have searched but the treasure has never been found.

The search for Abraham's gold continues to this day. If you find yourself out walking along the banks of Pheasant Branch near Flat Rock, North Carolina, where Abraham's body was found, keep your eyes peeled for a twisted and ancient white oak tree. But watch out for Abraham's ghost, he's known to haunt the place, sending scavengers wandering lost in the wrong direction. You might see him in the moonlight, with his shovel in hand, digging desperately.

APPENDIX B

The Jacob Carroll Migration

The Settlements and Boundaries of Rowan County. (The University of North Carolina, the James Sprunt Historical Publications, Published under the direction of the North Carolina Historical Society, J.G. de Roulhac Hamilton, Editor & Henry McGilbert Wagstaff, Editor, Vol. 16, No. 1).

Jacob's and Dennis' grandfather, William Carroll, emigrated from Northern Ireland to Charleston, South Carolina in the early 1700s and settled in Rowan County, North Carolina. William Carroll was an early constable and commissioner of the county and died in Rowan County in 1754, leaving five minor children including Dennis Carroll Sr., Jacob's and Dennis Carroll, Jr.'s father.

William's grandchildren, Dennis Carroll, Jr. (1793 – 1859) and Jacob Carroll (1798 – 1880) were born in South Carolina. Dennis, Jacob and their families moved across the state line to Rutherford County, North Carolina, and then to Georgia and on to Arkansas. Dennis Carroll, Jr. and his family remained in Arkansas; Jacob Carroll and family continued on to Texas. Jacob and his wife Sallie are buried in Glaze County, Texas.

The first emigrant, William Carroll, settled on a land grant from the Earl of Granville. "The Earl Granville disposed of his lands in Carolina upon favorable terms, for he desired to increase their value by rapid settlement. Therefore, influenced by the inviting nature of the climate and soil, the peacefulness of the Catawba Indians and the laxity of North Carolina laws in comparison with those of Virginia on the subject of religion, the Scotch-Irish…made homes for themselves in western North Carolina."

"In the year 1729, the King of Great Britain, according to act of Parliament, purchased seven-eighths of the territory of the Carolinas from the Lords proprietors, for twenty-five hundred pounds for each eighth part. But John, Earl of Granville, the son and heir of Sir George Carteret, refused to part with his portion, and his lands were laid off to him, extending from latitude thirty-five degrees, thirty-four minutes on the Virginia line, and westward to the South Sea, or pacific Ocean. It is within the limits of Earl Granville's lands and on the western portion of them that Rowan County was situated.

"Here Earl Granville's lands – lately set off to him – were offered at a cheap rate, and the climate was much more mild than in the homes they had chosen in Pennsylvania. The first arrival of Germans in Western North

Carolina, in the bounds of Old Rowan, is believed to have taken place about 1745...They came side by side with their Scotch-Irish neighbors, sometimes settling in the same community with them.

"Old Rowan included in its ample domain the territory occupied today by thirty counties and parts of counties in North Carolina...Granville does not appear to have exercised any authority over the people in his lands, nor any control in the enactment or execution of the laws. He was simply a mighty landowner, with a vast body of desirable land to sell to the best advantage. After 1743 all grants and sales of lands were made in his name. The curious inquirer may look into the office of our Register of Deeds, in the Courthouse in Salisbury, and see volumes upon volumes of old land deeds."

APPENDIX C

The History of Rowan County, NC

Jethro Rumple, *A History of Rowan County, North Carolina*. (1916. Reprinted by Heritage Books, 2005. Full text available at The Internet Archive, https://archive.org/stream/historyofrowanco00rump/historyofrowanco00rump_djvu.txt, accessed 7/19/2014).

The Rowan County of William Carroll's time was described by Lawson, in: History of a Journey from Charleston to Pamlico Sound, in the year 1701, as a: "pleasant savanna ground, high and dry, having very few trees upon it, and those standing at a great distance; free from grubs or underwood. A man near Sapona may more easily clear ten acres of ground than in some places he can one; there being much loose stone upon the land, lying very convenient for making of dry walls or any other sort of durable fence. The country abounds likewise with curious bold creeks, navigable for small craft, disgorging themselves into the main rivers that vent themselves into the ocean…We reached the fertile and pleasant banks of the Sapona River, whereon stands the Indian town and fort; nor could all Europe afford a pleasanter stream, were it inhabited by Christians and cultivated by ingenious hands. This most pleasant river may be something louder than the Thames at Kingston, keeping a continual warbling noise with its reverberating upon the bright marble rocks…It is beautified by a numerous train of swans and other waterfowl, not common, though extraordinary pleasing to the eye. One side of the river is hemmed in with mountainy ground, the other side proving as rich a soil as any…" (The native Catawba and "Sugarees" [other name: Usherees] from Sugar Creek lived in the dense woods explored by Lawson by boat up the Santee River.)

Lawson, the author, describing the same area again in the late 1800's laments that, "the swans, beavers, deer, and buffaloes have fled before the march of civilization, but on the south side of the stream still stand the bold bluffs rising abruptly from the river bank. Some of these heights are now clothed with cedars and other forest trees, but one of them is crowned with an old family mansion, that was formerly known as 'The Heights of Gowerie'. At the foot of this is a spring of pure cold water, and nearby a mill, driven by water drawn from the river above by a long canal…Not many years ago a lady, with the hectic flush upon her cheeks, returned from a distant land to visit for the last time her native place—the old mansion on the hill. She was accompanied by a gentleman residing in the neighborhood, who after her departure penned the following lines:

HEIGHTS OF GOWERIE

Pensive I stand on Gowerie's height,
All bathed in autumn's mellow light –
   My childhood's happy home;
Where Yadkin rolls ints tide along
With many a wail and mournful song,
   As its waters dash and foam.
And memory's harp tunes all its strings,
When I catch the dirge the river sings,
   As it sweeps by Goweries' side.
And viewless tongues oft speak to me,
Some in sorrow and some in glee,
   From the river's fitful tide.
On yon isle, just up the river,
Where sunbeams dance and leaflets quiver,
   Three fancied forms I see.
That blest—that sainted trio band,
Together walk adown the strand.
   And wave their hands at me.
A father 'tis, whom yet I mourn.
And sisters two, who long have gone –
   Gone to the other shore.
They beck me to the goodly land.
Where, with them, I'll walk hand in hand,
   Ne'er to be parted more.
When from the found hard by the mill,
Just at the foot of Gowerie's hill,
   I drink the sparkling water;
Echoes from yon cedar grove,
From which the sighing zephyrs rove,
   Say, "come to me, my daughter."'

APPENDIX D

Peter Carroll

From the Sheridan Press, Oct. 3, 1911

"Favors Mr. Carroll"

"Editor Post: I see in your issue of September 26th the question asked, 'Why do not the taxpayers and electors of the city select whom they prefer to have represent them on the board of city commissioners?'

As you say, it is well for the honest citizens of the city to stop and think who it was that stood by them in the late controversy between themselves and a big corporation.

There is an old saying that it is well to 'speak well of the bridge that carries you safely over.' I think this will apply. In this case, to G.G. Carroll, one of the members of the present city council; and we of the north end are willing and anxious for him to be one of the three who will guide the municipal ship under the new system. We know that he will stand firm in any storm.

Now Mr. Carroll, it is up to you to let your friends know whether you will accept the place." A Taxpayer

From the Sheridan Press, Oct. 27, 1911

"Highly Honored"

I want to express my thanks to the people of Sheridan for their support at the primary, Tuesday. I accept this vote as an endorsement of the course I have pursued while a member of the city council, and feel highly honored by this expression of confidence in me.

If I am elected commissioner at the November 7th election, I can only say that I will continue to serve the people to the very best of my ability, and knowing the will of those who honor me with their votes, I will do my best to carry it out, and will make it my business to see that the interests of Sheridan are carefully guarded and that every person in the city gets a square deal.

From the Sheridan Press, 1949

Peter Carroll, Early Sheridan County Legislator, Dies Here

Peter Carroll, a resident of Sheridan since 1901, died early this morning at Memorial hospital where he had been a patient two days. He was 88 years of age.

Carroll was born June 8, 1860, in Scotland and came to the United Stated when a small child. His family settled first in Pennsylvania and later moved to Iowa.

On moving to Sheridan at the turn of the century, Carroll went to work in the mines at Dietz. He was active at that time in labor circles and later when he was elected to the state house of representatives in the legislature, was instrumental in the passage of the eight-hour labor law.

Carroll served in the legislature for three terms. He held life membership in the Knights of Pythias and was one of the oldest members of the First Baptist church. For the past several years, he had made his home with his son, Jim Carroll.

APPENDIX E

George G. Carroll

From the Sheridan Press, 1911

"Under the Commission Form"

"In 1892, seeing the possibilities of wealth in the cattle business, Mr. Carroll sold his ranch and moved with his family to the Rosebud country, where he purchased a squatters right and began life as a cattle man. At this business he prospered. His cattle increased in numbers and gradually he accumulated land until in 11 years he was the owner of some 1,500 well bred cattle and 550 acres of land.

Then came the grasshoppers and adversity. The first year the hoppers took the feed and the range was as bare as a bone. Most of the ranchers sold their stock for what they could get, but George held on. For three years he hoped against hope, but each year the hoppers took it all.

Finally he shipped his cattle to Nebraska, and later trailed them back to the range. To those who know the country through which these cattle had to be trailed the result of the venture need not be told. Sufficient to say, by the time Mr. Carroll had cleaned up his herd, he was mighty near cleaned up himself. Still he saved something out of the wreck, and with it established himself in business in Sheridan.

For a time he owned a livery barn located on Smith Street, and as the man he had employed as manager was accidentally killed, Mr. Carroll took over the management personally and conducted the barn until he could dispose of it. Three years ago last spring he became part owner in the big flowering mill west of Big Goose, and operated it until a few months ago.

A year ago he was elected councilman, and when the commission plan of government was adopted became a candidate for commissioner. There were a score of candidates, but Mr. Carroll was nominated and when election came he polled the largest vote of any of the four candidates for the office of commissioner. The only office he ever held was that of councilman and his record in that position is one of which he has good cause to be proud. It cannot and will not be denied that upon that record Mr. Carroll was elected to the position he will occupy after January first, the only of the eight councilmen to be so honored.

George Carroll does not claim to be a gifted statesman nor an infallible legislator, but he will do what he believes is right, and the people know it. He is neither bullheaded nor is he easily led, but looks at every proposition carefully, does not hesitate to consult with others who might be better posted on certain phases of the question than he is , then he makes up his mind, and having done that, he is there to stay. He is neither loud-mouthed nor is he domineering in his discussions of public questions, but whenever a question arises in which the interests of the people are at stake, you do not need a spy glass to find George Carroll, for he is always there and he brings the goods with him.

APPENDIX F

COFFIN FAMILY PAPERS, 1864-1977

Washington State University Libraries
Manuscripts, Archives, and Special Collections
Pullman, WA 99164-5610 USA
(509) 335-6691
Cage 405
The papers of the H. Stanley Coffin Family were donated to the Washington State University Libraries by H. Stanley Coffin III in July 1977 (77-39). The collection was processed from June to November 1979 by Pat Graham Pidcock.
Number of containers 15
Linear feet of shelf space 19
Approximate number of items 12,340

Personal and Business History:

"For over fifty years, the firm of Coffin Brothers, Incorporated, dominated the livestock industry in central Washington, the wholesale grocery business in the Yakima Valley, and a variety of other financial enterprises. Coffin Brothers became the core of a family owned group of business ventures. The principal founder of the family fortunes was H. (Harvard) Stanley Coffin, born at The Dalles, Oregon, on September 14, 1869. He attended local public school, the Wasco Academy, and a business college. At age fourteen he began working in the warehouse of McFarland and French, General Merchandisers. Four years later, he moved up the Columbia River to Arlington, Oregon, where he worked with his brother, Arthur Coffin at the Coffin-McFarland Company warehouse. Stanley and another brother, Lester, learned the business quickly, and proceeded to purchase a quarter interest in the firm which was then renamed Arthur Coffin and Brothers...

"Over the years the Coffin Family controlled substantial real estate holdings. Before 1909, they had acquired 50,000 deeded acres and 50,000 leased acres between Ellensburg and Wenatchee, known as the Cape Horn/Mountain Home Ranch. Eventually, Lester Coffin's descendents came to manage this property. Stanley Coffin I took control of the Coffin/Babcock range (also called the West Bar Ranch) which was located on the Columbia River. The family's Wenas Creek lands included the Mt.

Vale Ranch, the McCabe Place, and the Lee Land ranch which totaled approximately 16,000 acres. Lying west of Yakima, the Cold Creek Ranch contained about 17,000 acres of well-watered grazing land. The Horse Heaven Ranch, inclusive of the Wilmert place and the Blakely place, was located south of Kennewick and was comprised of 30,000 deeded acres plus adjacent leased pastures. In addition to small deeded holdings (i.e., city lots, beach plots) the family also controlled various tracts of leased acreage. . . ."

APPENDIX G

STANDARD OPERATING PROCEDURE FOR THE JUMP AND SECURING EQUIPMENT

June 16, 1943

Headquarters, 501st Parachute Infantry Camp Mackall, N.C.
Training Memorandum numbered . . . 27 (Excerpted below:)

The following instructions constitute standard operating procedure for all men of this organization making jumps during unit training. These instructions supersede all previous instructions in contradiction thereto and will be followed to the letter.

PART I.
JUMPMASTER INSPECTION PRIOR TO EMPLANING

A. Jumpmaster inspection of personnel and their equipment – (See Annex "A", attached.)
B. Inspection of delivery units and equipment chutes – (See Annex "A", attached.)
C. Inspection of the plane – (See Annex "A", attached.)

PART II
JUMPMASTER PILOT CONFERENCE

This conference will include complete discussion between pilot and jumpmaster of type of jumping, initial flight pattern, formation in the air, direction and velocity of surface winds and winds aloft, approach to the jump field, including review of navigational check points, altitude of plane prior to the approach log, and altitude and type of approach desired, and pilot signals to include the pilot's alert signal (horn or bell), the stand-to-door signal (red light) and the "go" signal (green light) and speed of plane during jump. For standard flight formation, jumping altitude of planes and pilot signals, see Training Memorandum No. 4, Hq Airborne Command, May 8, 1943.

PART III
LOADING OF PERSONNEL AND EQUIPMENT

The jumpmaster will personally supervise the loading of equipment bundles and personnel into the plane. Where para-packs are to be used, the jumpmaster will request assistance from the crew chief in loading the bomb racks. The supervision of loading will include completion in triplicate of a jumpmaster form attached hereto see Annex "B". One such form will be given crew chief, one sent to Regimental S-3, and one to unit jumping.

Jumpmaster will require that personnel enter the plane in proper inverted order and seat themselves as far forward as possible, successive men to sit on alternate sides of the walk-way. When men are seated, jumpmaster will require that they hook safety belts and wear helmets until after the plane has taken off. Equipment bundles will be so positioned as to (a) not obstruct the door or walkway; (b) not to be in or protruding from the door; (c) to be as near the door as safe loading of the plane and non-obstruction of the door permits.

Having supervised the loading of personnel and equipment, jumpmaster will notify crew chief that the parachute personnel and equipment is ready for take-off; whereupon jumpmaster will spend time prior to take-off issuing any additional instructions to pilot or parachute personnel and will himself be seated and hook his safety belt prior to take-off. After take-off when plane is safely in flight the jumpmaster will command "Unhook Safety Belts!", position himself either at the door or with the pilot, wherever he can best assist the pilot in location of the objective jumping area and placing plane on proper final bearing. At least five minutes prior to estimated time of the alert signal, jumpmaster will require that bundles be moved to such a position that outer-most edge of the bundle is against the right side of the door and flush with the bottom of the door, and with bundle snap fastener hooked up with only sufficient static line out to allow no loose line upon the floor. See sketch "I," Annex "C."

On sounding of the alert signal, the jumpmaster will commence his check of equipment and will command, "Stand Up!", "Hook Up!", "Check Equipment!", and will himself then make a rigid inspection of each man, see Annex "A." When he is satisfied that all men are ready to jump and are correctly booked up, he will command, "Sound off for equipment check!" When all men have signified equipment "O.K", he will himself hook up as No 1 man of the stick with his snap fastener next to the equipment bundle. This should be completed prior to the 2 minute red light signal. Then, with the aid of No. 2 he will slide the bundle to a position roughly one-third outside the edge of the door (see Sketch No. 2, Annex C) and will begin checking his position over the ground and attempt to pick up the on-coming objective dropping area. When the stand-to-door signal is given by the pilot, the jumpmaster with the aid of No. 2, if necessary, will shove the bundle farther out of the door to such a position that the bundle is just under one-half the way out the door; i.e.; to a point where only a small additional push is required to eject the bundle from the plane. The jumpmaster will then kneel on the end of the bundle with his head near the right side of the door to pick up the drop area and await the pilot's green light, "Go" signal.

No. 2 man, who will have been so instructed by the jumpmaster, will be observing the pilot's light signals from the time when the red light goes on, except for such time as he may be required to assist No. 1. **When No. 2** observes the green light to flash on, he will strike the jumpmaster a smart blow on the left shoulder with his fist. [Emphasis added.]

THE EXIT: When the jumpmaster has received No. 2's blow upon his left shoulder, he will (precluding a condition in the plane which would render the jump unsafe) eject the bundle and follow it a rapidly as proper body position will permit, (if more than one bundle is ejected at a time, the jumpmaster may require the assistance of No.2. In this case, No. 1 will position himself at the left rear of the bundles and No 2 at the right rear. CAUTION: No. 1 must be particularly careful to be on the right side of the bundles static line.)

The Jumpmaster will be followed in turn by all members of the stick at such a pace as will permit each man to obtain a good position on exit. This control type or individual exit type does not preclude rapid exit but is as rapid as the mad rush type of exit and has the obvious advantage of greater safety for each man.

THE DESCENT: The jumpmaster, as well as every other man in the stick, will observe the five points of performance, which are (1) checking body position and counting on exit, (2) checking canopy, (3) keeping a sharp look out during descent, (4) preparing to land, (5) landing. Each man will also orient himself with respect to (1) the ground and in particular the location of his assembly area, and (20 the position at which the equipment bundle will be located.

THE ASSEMBLY ON THE GROUND: Immediately on landing, each man from the prone position will remove his harness in this order, left snap of reserve, belly band, left leg strap, right leg strap, and secure his M-1, carbine or sub-machine gun from its container as rapidly as possible, meanwhile observing about him to determine the proper direction to the equipment bundle and his location with respect to the assembly area, kneeling if tactical situation so dictates.

Having removed his harness and obtained his individual weapon, each man will begin working toward his equipment bundle as rapidly as the tactical situation permits. The first man to arrive at his proper bundle will not open the bundle but will observe for members of his squad to ascertain that at least one other member of his unit is on his way to the equipment bundle; the first man will begin to pen the bundle as rapidly as possible. The second man to arrive in the vicinity of the bundle will likewise not go to the bundle until he is certain that at least a third member of the unit is heading toward the bundle. Having discerned a third member approaching the bundle, No. 2 will then go to the bundle and assist No 1 in opening it and securing equipment therefrom. As subsequent members of the squad approach the bundle, they will remain dispersed in the vicinity of the bundle and will signify their direction with respect to the bundle or the squad leader, if on hand, will then direct individuals to approach the bundle singly, or in pairs, or larger groups as the tactical situation dictates. The man directing this securing of equipment will likewise designate the position to which members of the squad will go immediately after having obtained their equipment. At this position the squad leader or senior man will take charge and direct the movement of the unit to the assembly area by such method as the tactical situation shall dictate. This ends the phase of the tactical problem, which employs the plane and chute as a means of transport to the scene of ground action.

APPENDIX H

DESCENDANCY OF PETER CARROLL & MARTHA ATHERTON *and* GEORGE G. CARROLL & MARY ELLA KIRBY

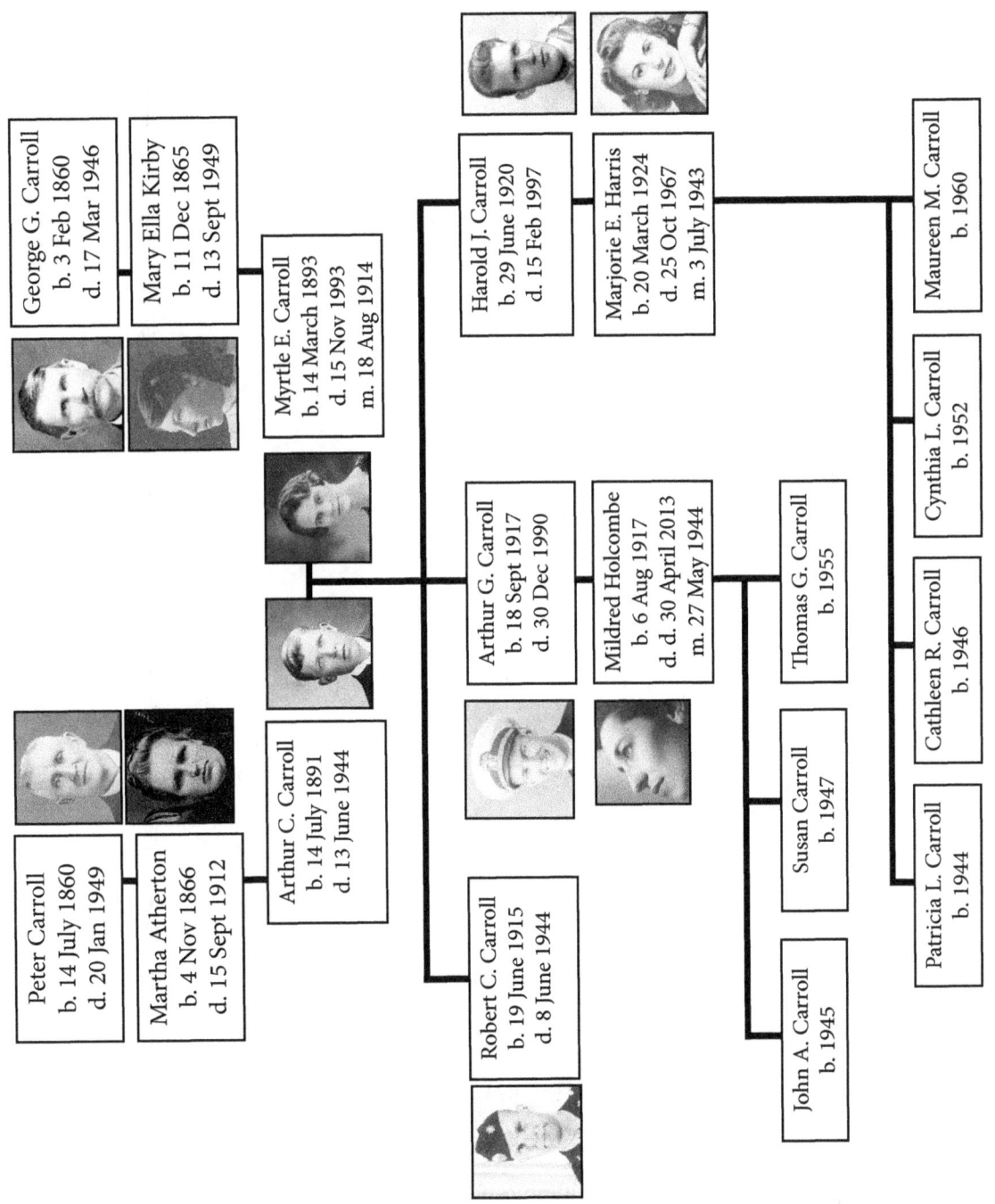

APPENDIX I

GEORGE W. CARROLL FAMILY TREE

Name	GEORGE WASHINGTON CARROLL				Information on this sheet obtained from				
Born	ca 1900	Where	Atlanta, Georgia						
Chr.		Where	Dayton, Wyoming		**GEORGE WASHINGTON CARROLL** (Husband's Full Name)				
Mar.		Where			**NANCY ABBIGAIL MC ELROY** (Wife's Maiden Name)				
Died		Where			Date: July 27, 1966				
Father: Jacob Carroll		His Mother's Maiden Name			Compiler: Merle Harris				
Wife's Name: Nancy Abbigail McElroy					Address: 538 77th Street N. E.				
Born 26 July 1837		Where Macon, Georgia (per Death Certificate)			City: Cedar Rapids State: Iowa				
Died 11 Oct. 1925		Where Fromberg, Carbon County, Montana							
Father: John McElroy		Her Mother's Maiden Name (?) Kirkindol							

CHILDREN Arrange in order of birth	When Born Day Month Year	Where Born Town or Place	County	State or Country	Married to	When Married Day Month Year	When Died Day Month Year	Where Buried Town or Place	County
George Granville	3 Feb. 1860	Gonzales	Gonzales	Texas	Mary Ella Kirby	21 Jan. 1886	17 Mar. 1946	Sheridan	Sheridan Wy
					Mat Jones			Fromberg	
William	24 Feb. 1865	Belton	Bell	Texas	Irvin Russell		10 Oct. 1950	Bur. Rockvale, Cha.Mo	
Mollie					J. D. Thorn			Long Beach	Ca
Fred								Sheridan	Sheridan Wy
Solon								Sheridan	Sheridan Wy

APPENDIX J

GEORGE GRANVILLE CARROLL FAMILY TREE

Name: GEORGE GRANVILLE CARROLL
Born: 3 Feb. 1860 — Where: 13 miles south of Gonzales, Gonzales County, Texas

17 March 1946	Where: Sheridan, Wyoming
19 March 1946	Where: Municipal Cemetery, Sheridan, Wyo.
21 Jan. 1886	Where: Young County, Texas

Husband: George Washington Carroll — His Mother's Maiden Name: Nancy Abigail McElroy

Wife: Mary Ella Kirby
11 Dec. 1865 — Where: McLean County, Ill.
23 Sept. 1939 — Where: Sheridan, Wyo.
25 Sept. 1939 — Where: Sheridan, Wyo.

Father: John M. Kirby — Her Mother's Maiden Name: Rachel Sells (Rachel Sells)

Husband's Full Name: GEORGE GRANVILLE CARROLL
Wife's Maiden Name: MARY ELLA KIRBY

Date: January 26, 196_
Compiler: Merle Harris
Address: 538 27th St. N.
City: Cedar Rapids

CHILDREN

Name	Day	Month	Year	Town or Place	County	State	Married to	Day	Month	Year
Granville W.	17	Feb.	1889	Dayton	Sheridan	Wyo.	Alta Wallace			
Myrtle Ella	14	Mar.	1893	Dayton	Sheridan	Wyo.	Arthur Colin Carroll	18	Aug.	1914
							Frank McEowen	13	July	1947
Jewell Ethel	2	July	1895	Kirby	Rosebud	Mont.	Roy Allen			
							Thomas W. Reed			
Enola Vera	12	Jan.	1898	Kirby	Rosebud	Mont.	James Phillips			
Orpha Rachel	11	Aug.	1899	Kirby	Rosebud	Mont.	H. Buford Walker	4	Mar.	1928

APPENDIX K

PETER CARROLL FAMILY TREE

PETER CARROLL (Husband's Full Name)
- June 3, 1860 — Where: Glasgow, Scotland
- Jan. 20, 1949 — Where: Kalo, Iowa
- Jan. 20, 1949 — Where: Sheridan, Wyoming
- March 25, 1885 — Where: Kalo, Webster County, Iowa

James Carroll — His Mother's Maiden Name: Catherine Sharp

NAME: Martha Atherton (Wife's Maiden Name)
- Nov. 4, 1866 — Where: Syracuse, New York
- 1888 — Where: Kalo, Iowa
- Sept. 15, 1912 — Where: Sheridan, Wyoming
- Sept. 17, 1912 — Where: Sheridan, Wyoming

Edward James Atherton — Her Mother's Maiden Name: Sarah Simpson

Information on this sheet obtained from:
Compiler: Myrle Harris
Address: 538 27th Street N. E.
City: Cedar Rapids State: Iowa 5402
Date: July 25, 1966

CHILDREN (Arrange in order as born)	When Born Day Month Year	Where Born Town or Place	County	State or Country	Married to	When Married Day Month Year	When Died Day Month Year	Where Buried Town or Place	County	State or Country
Edward J.	21 Nov. 1885	Kalo	Webster	Iowa			10 Aug. 1887	Otho	Webster	Iowa
Myrtle L.	8 Nov. 1887	Boone	Boone	Iowa	James J. Bledsoe	4 Sep. 1906				
Robert E.	9 Oct. 1889	Boone	Boone	Iowa	Vina Watson	4 Nov. 1908	7 July 1966	Sheridan	Sheridan	Wyo.
Arthur Colin	14 July 1891	Boone	Boone	Iowa	Myrtle Elis Carroll	18 Aug. 1914	13 June 1944	Sheridan	Sheridan	Wyo.
Margaret Ruth	12 Aug. 1893	Ft. Dodge	Webster	Iowa	Marlowe Ferris	Dec. 1915				
Estella M.	23 June 1895	Ft. Dodge	Webster	Iowa			12 Aug. 1895	Otho	Webster	Iowa
Lawrence R.	28 Jan. 1897	Ft. Dodge	Webster	Iowa			1 Apr. 1904	Sheridan	Sheridan	Wyo.
James E.	23 July 1900	Red Lodge	Carbon	Mont.	Rosalie Siagle	1923	17 Apr. 1960	Sheridan	Sheridan	Wyo.
Paul	15 Mar. 1904	Dietz		Wyo.			26 Mar. 1904	Sheridan	Sheridan	Wyo.

APPENDIX L

JOHN M. KIRBY FAMILY TREE

Name: JOHN M. KIRBY
Born: July 22, 1900 — Where: Paola, Kansas
Died: Nov. 18, 1860
His Father: Jacob Kirby — His Mother's Maiden Name:
Her Name: Rachael Sells — Where: Dixon, Lee County, Illinois
Married: Oct. 21, 1887 — Where: Graham, Young County, Texas

Her Father: Anthony Sells — Her Mother's Maiden Name: Esther Mary Nundell

Husband's Full Name: JOHN M. KIRBY
Wife's Maiden Name: RACHAEL SELLS
Date: July 27, 1966
Compiler: Merle Harris
Address: 538 27th Street N. E.
City: Cedar Rapids — State: Iowa

CHILDREN (Arrange in order of birth)	When Born Day Month Year	Where Born Town or Place	Where Born County	State or Country	Married to	When Married Day Month Year	When Died Day Month Year	Where Buried Town or Place
James Albert	23 Feb. 1862	Bloomington	Woodford	Ill.	Ruby Virginia Payne	Dec. 1888	12 Apr. 1938	Lubbock
George E.	16 Nov. 1863	Fairbury		Ill.	Ada L. Keller		12 May 1945	Billings Yellowstone
Mary Elta	11 Dec. 1865	Fairbury		Ill.	George Granville Carroll	21 Jan. 1886	23 Sept. 1939	Sheridan Sheridan
Orpha		1867			Wesley Gregg		31 May 1890	
John Ira	11 Aug. 1869		Woodford	Ill.	Never married		12 Jan. 1943	Sheridan Sheridan
William Buckingham	29 June 1871				Linnie Hawkins		1964	Long Beach
Emma	28 Dec. 1872				Adelbert Church		4 Nov. 1939	Sheridan Sheridan
Dora	Dec. 1875				Clifford Austin		7 Mar. 1963	Klamath Falls
Charles R.	1879				Denis		1962	Eugene

NOTES

PART I, Chapter 4

1. Merle Harris, genealogist; research from Forsyth County, Georgia records, 'Hightower' is the Anglicized version of 'Etowah,' the Cherokee name for the area. There were no 'high towers' as I envisioned when I first read this fact.

2. Tennessee Ernie Ford – The Ballad of Davy Crockett lyrics. (http://www.songlyrics.com/tennessee-ernie-ford/the-ballad-of-davy-crockett-lyrics/, accessed May 15, 2014).

3. Emmet Starr, *History of the Cherokee Indians and Their Legends and Folk Lore* (The Warden Co., 1922). "A large body of Cherokees volunteered to assist the army led by Generals Andrew Jackson and John Colfie. In the crucial battle of Horse Shoe Bend in which the Creeks were strongly barricaded behind cypress log ramparts and were holding their own against the frontal attacks, a detachment of Cherokees came up on the opposite side of the river, Charles Reese swam across and towed a canoe to his associates, the canoe load of warriors landed in the back part of the bend, attacked the Creeks from the rear. In attempting to repel this assault the Creeks so weakened their front that a breach was made nearly annihilating the belligerent Creek forces. From that day Andrew Jackson became increasingly popular."

4. The Texas State Historical Association. (http://www.tshaonline.org/handbook/online/articles/fcr24, accessed: 3/29/2014). "Crockett died in the battle of the Alamo on March 6, 1836. The manner of his death was uncertain, however, until the publication in 1975 of the diary of Lt. José Enrique de la Peña. Susanna Dickinson, wife of Almaron Dickinson, an officer at the Alamo, said Crockett died on the outside, one of the earliest to fall. Joe, Travis' slave and the only male Texan to survive the battle, reported seeing Crockett lying dead with slain Mexicans around him and stated that only one man, named Warner, surrendered to the Mexicans. (Warner was taken to Santa Anna and promptly shot). When Peña's eyewitness account was placed together with other corroborating documents, Crockett's central part in the defense became clear. Travis had previously written that during the first bombardment Crockett was everywhere in the Alamo 'animating the men to do their duty.' Other reports told of the deadly fire of his rifle that killed five Mexican gunners in succession, as they each attempted to fire a cannon bearing on the fort, and that he may have just missed Santa Anna, who thought himself out of range of all the defenders' rifles. Crockett and five or six others were captured when the Mexican troops took the Alamo at about six o'clock that morning, even though Santa Anna had ordered that no prisoners be taken. The general, infuriated when some of his officers brought the Americans before him to try to intercede for their lives, ordered them executed immediately. They were bayoneted and then shot. Crockett's reputation and that of the other survivors was not, as some have suggested, sullied by their capture. Their dignity and bravery was, in fact, further underscored by Peña's recounting that 'these unfortunates died without

complaining and without humiliating themselves before their torturers."'

5. Wallace L. McKeehan, "DeWitt Colony Papers, Documents and Correspondence, 1824 – 1826." (http://www.tamu.edu/faculty/ccbn/dewitt/papers24-26.htm. Sons of DeWitt Colony Texas, http://www.tamu.edu/faculty/ccbn/dewitt/dewitt.htm. Wallace L. McKeehan, ed., accessed 3/28/14).

6. Ibid.

John McMullen and James McGloin settled families from Ireland and Mexico north of the Nueces; their settlement would later be known as San Patricio. They had great difficulty reaching their destination and settled temporarily in an abandoned Spanish mission, Mission Refugio.

The Mission was a large stone church, truly a refuge from the harsh sun, the cold night, the relentless wind and hostile Indians. It had served the Spanish missionaries and it also protected around 35 frightened and hungry Irish who had arrived at Port El Copano (near Corpus Christy) with high hopes. They now huddled in this lonely desert church, hopeless.

The Nueces River region was home to the Lipan Apache but it was the Comanche who raided – stealing horses and frightening the women and children behind the walls of the church where they stayed, leaving only for water. Empresario McMullen told the Mexican officials they couldn't go forward without supplies from New Orleans and Goliad, and that they needed backup desperately. But the Mexican government ordered them to evacuate the Mission and the miserable party continued upriver on their own resources.

Stories were told of the bright sun breaking through the clouds just when they reached a particularly scenic rise overlooking the Nueces River. Smiles of gratitude passed between them. Father Doyle, their Irish Priest, christened the imagined town-to-be San Patricio de Hibernia in the memory of their beloved Irish homeland.

Things went poorly from the beginning. While their women were still in mourning clothes over family and friends who'd died trying to reach this dangerous place, many of the men felt they'd been swindled. Many colonists pressed Padre Doyle to talk to the Mexican officials on their behalf. Several asked for transportation to New Orleans where they could make a new start or try to gain passage somewhere else; anywhere but Tejas! Some left on their own; some colonists stayed and built homes.

James Power and James Hewetson, natives of Ireland, settled Irish and a few Mexican families beginning in 1828 along the Nueces in the area between the Guadalupe and the Lavaca Rivers, eventually known as the Refugio colony. Power had been encouraged to bring Irish to Tejas by one of the earliest settlers along the Guadalupe, Captain Portilla. In 1808, at the request of the Spanish government who then controlled the land, Portilla led sixteen Spanish and Mexican families to settle where the Camino Real crosses the Guadalupe River just north of what would become Gonzales. When they left in 1812, due to never ending Indian attacks and raids, the retreating settlers failed to gather all of their stray cattle to the later joy of Texas settlers who rounded up their descendants. Captain Portilla and his family eventually settled along the Rio Grande and remained influential in the area.

San Patricio's citizens were treated to a rare celebration in July, 1832, when James Power married Captain Portilla's daughter, Dolores. The wedding lasted several days and brought folks from far and wide, including Captain Don Carlos de la Garza, who rode from his ranch south of La Bahia with a large group of ranch-hands. Known for its grand hacienda, the Carlos Ranch was a small town, in fact, with lots of horses and cattle, a chapel, a general store, barrel house and commissary, barns, sheds, and corrals. The Captain and his lady had sheltered many families from hostiles, outlaws and storms.

While San Patricio was thriving in 1834, the Rufugio colony was hard hit by the cholera epidemic which killed from one third to one-half of the two hundred and fifty colonists.

Haden Edwards also brought American families to the Nacogdoches area. Of disreputable character, he boldly tried to shake down the established Mexican and Anglo families, many who'd inhabited their land since the 1780's, for money and goods. These folks were not pushovers and Edwards was no genius. However, the Mexican government considered all current inhabitants to be "colonists" of one of the empresarios; Edwards required them to show proof of land ownership or pay! Most had never received titles from the Spanish government. He was eventually run-out-of Texas but his brother, Benjamin Edwards, with a band of Americans, kidnapped the local official, an Anglo old-timer, and held kangaroo court. They unsuccessfully declared independence from Mexico and announced the beginning of the Republic of Fredonia.

Other shady "Empresarios" sold their contracts to New York and Boston speculators never planning to move to Texas themselves. Robert Leftwich contracted to settle American families in the Brazos valley. His sole purpose was to acquire land for a group of Tennessee investors he represented called the Texas Association. About seventy stockholders bought land in the upper Brazos, but Leftwich failed to send any settlers until October 1830, after the Mexican government had passed a law restricting further immigration from the United States. The nine families sent were detained but eventually allowed to settle.

7. Ibid. "In 1828, General Manuel Mier y Terán was commissioned by President Guadalupe Victoria to help in negotiating the boundary between the Republic of Mexico and the United States of the north and determine the situation in the colonies in more detail after the Fredonian Rebellion. He was the head of a scientific commission spending most of the time in Nacogdoches in 1828 gathering data for a boundary survey."

8. Texas State Historical Association, "Creed Taylor." (http://www.tshaonline.org/handbook/online/articles/fta17, accessed 3/29/2014).

Chapter 5

9. Jacob Van Zandt (1751-1818) appears on the Rutherford Co. NC 1782 Tax List, being assessed £180 on 400 acres, 9 horses and 18 cattle. He did not collect a pension, nor did his wife Catherine Moon or any of his heirs. See appendix for more information about the Texas Van Zandts.

10. Abraham Kuykendal Sr. (1719 – 1812) had a son and Jacob Van Zandt (1751 – 1818) had a daughter who married: Elizabeth Van Zandt (1772 – 1829) married Abraham Kuykendall, Jr. (1770 – 1871). Their daughter, Elanor Van Zandt Kuykendal (1809 – about 1860), married William McElroy (1809 – 1862). Elanor and William McElroy's daughter was Nancy Abigail (Abby) McElroy Carroll.

11. Ancestry.com, Dennis Carroll Jr. and Nancy Waggoner's family tree. (http://www.ancestry.com/, accessed 7/18/2014).

Chapter 6

12. Mystery surrounds Jacob Carroll's wife, Sarah "Sallie" Carroll, including her unknown parentage. Perhaps she was of Cherokee or Melungean descent. There is no explanation given in the church records, or elsewhere, for Jacob's excommunication from the Mt. Tabor Baptist Church. The two, Sallie's unknown or unacknowledged heritage, and Jacob's excommunication might be interrelated. Further, prior to the "trail of tears," many white settlers intermarried with Cherokee, afterward, a family members genetic history might be secreted.

13. Mt. Tabor Baptist Church, 1800's Membership Lists of Mt. Tabor Baptist Church, Georgia.

14. Garland C. Bagley, *History of Forsythe County GA, 1832 – 1932* (Boyd Publishing Company, 1997).

15. Ancestry.com, Dennis Carroll Jr. and Nancy Waggoner's family tree. (http://www.ancestry.com/, accessed 7/18/2014).

16. Michael A. Lofaro, "Crockett, David," Handbook of Texas Online (Texas State Historical Association, uploaded 6/12/2010, modified 1/27/2014; http://www.tshaonline.org/handbook/online/articles/fcr24, accessed 7/18/2014).

17. Jacob Crockett Carroll (1821-1890) was named after Davy Crockett. He was Jacob Carroll's nephew, a son of Dennis Carroll Jr.

18. Thomas Nuttall, *A Journal of Travels into the Arkansas Territory During the Year 1819* (University of Oklahoma Press, 1980).

19. Term coined by T.H. Breen; quoted by Joshua Lee McKaughan in Journal of Backcountry Studies article "Barcelonia Neckerchiefs, Teaware, and China Plates: Kinship Status and the Division of Fourth Creek Church." (http://www.partnershipsjournal.org/index.php/jbc, accessed 3/29/2014).

20. Joshua Lee McKaughan in Journal of Backcountry Studies article "Barcelonia Neckerchiefs, Teaware, and China Plates: Kinship Status and the Division of Fourth Creek Church." http://www.partnershipsjournal.org/index.php/jbc. Date of Access: 3/29/2014. Imported British cups were a status symbol. Being descendant from one of the first families in the Colonies was also prestigious. Tea sets were passed down in the Carroll family. I am in possession of a labeled cup and saucer: black background with gold edging and a design of gold leaves and flowers. The cup and saucer are stamped Royal Grafton Fine Bone China Made in England and they are marked: "10" and "1147". The label (tape) reads: Emma Meide, Edna Srite, Valberg Hull/ - N.C., Henrietta Stephens, Ardis Dixon, District Press 1966.

21. Ancestry.com, "De Witt's Colony," citing Kuykendall, *A Texas Scrap Book*, 145. The Karankawa were thought to be cannibals, yet according to Kuykendall, "The only cannibalism to which they were addicted was that of eating pieces of an enemy's flesh at a war dance to inspire them with courage."

22. Daniel Shipman, *Frontier Life: 56 Years in Texas* (1879), Texas State Historical Association, (http://www.tshaonline.org/handbook/online/articles/fsh30, accessed 3/29/2014). The Kuykendall families were neighbors and friends of the Shipman family as shown by the fact that in January, 1771, Daniel Shipman, Sr., the father of Sarah's husband, Jacob, witnessed a deed where the buyer was named as Abraham Kuykendall.

Chapter 7

23. *The History of Gonzales County, Texas* (Gonzales County Historical Commission, 1986).

24. Louis and Jake Jr. are also buried in Glaze City, TX. They died in 1870. Other reported headstones (I have not visited the site): Jacob Carroll 1798-1880, Sarah Carroll 1799 – 1882, James Turk 18??-1888, Sallie Turk 1840-1882. There are also slaves of the Carrolls buried here. The 1856 census lists York and Jack as members of the Baptist church and property of Jacob Carroll.

25. W.T. Block, *A History of Jefferson County, Texas[,] From Wilderness To Reconstruction* (Nederland Publishing Company, 1976), Chapter XII, "Early Religion, Education, and Social Interaction." (http://www.wtblock.com/wtblockjr/history%20of%20jefferson%20county/Chapter%2012.htm, accessed 7/18/2014).

26. *In Motion: The African American Migration Experience* (The New York Public Library, 2005), "Haitian Immigration: 18th & 19th Centuries—The Black Republic and Louisiana." (http://www.inmotionaame.org/migrations/topic.cfm?migration=5&topic=3, accessed 7/18/2014). After the trans-Atlantic slave trade had ended, the Haitian Revolution in 1804 created the first republic of free Africans on Haiti and sparked a mass exodus of refugees to New Orleans. "The 1809 migration brought 2,731 whites; 3,102 free persons of African descent; and 3,226 enslaved refugees to the city, doubling its French-speaking population."

27. Vada Campbell, "Montgomery County, Our Heritage" (unpublished). (The author continues her search for the Carroll library.)

28. Randolph B. Campbell, *Wealth and Power in Antebellum Texas* (Texas A&M University Press, 1977).

29. Ibid.

30. Robert Service, "The Cremation of Sam McGee" (1907).

31. Richard W. Stewart, ed., *American Military History, Volume II: The United States Army in a Global Era, 1917 – 2003* (United States Army, 2005), (http://www.history.army.mil/books/AMH-V2/AMH%20V2/chapter2.htm#b6, accessed 7/19/14). The CMTC "provided about 30,000 young volunteers with four weeks of military training in summer camps each year between 1921 and 1941. Those who completed three, later four, years of CMTC training and related home-study courses became eligible for commissions in the Officers' Reserve Corps.

The CMTC thus provided another source of leadership for the Organized Reserves. Although relatively few officers emerged directly from the program, a substantial number of CMTC participants later attended West Point, entered ROTC programs, or received commissions during World War II."

Chapter 8

32. Gilbert M. Cuthbertson, "Regulator-Moderator War," Handbook of Texas Online (http://www.tshaonline.org/handbook/online/articles/jcr01, accessed 7/18/2014. Uploaded 6/15/2010. Published by the Texas State Historical Association). Lasting from 1839 to 1844, this war "involved several hundred men on each side, and caused much bloodshed and violence."

33. Campbell, *Wealth and Power in Antebellum Texas*. "The expansion of slavery correlated closely with soaring cotton production, which rose from fewer than 60,000 bales in 1850 to more than 400,000 in 1860. Only a minority of antebellum Texans, however, actually owned slaves and participated directly in the cash-crop economy. Only one family in four held so much as a single slave, and more than half of those had fewer than five bondsmen. Small and large planters, defined respectively as those owning ten to nineteen and twenty or more slaves, held well over half of the state's slaves in both 1850 and 1860. This planter class profited from investments in land, labor, and cotton and, although a decided minority even among slaveholders, provided the driving force behind the state's economy."

34. Richard W. Slatta, *The Cowboy Encyclopedia* (WW Norton & Company, 1994).

35. Block, *A History of Jefferson County, Texas*.

36. Bruce Stanton, *The Regulator Moderator War*, (http://www.kilgorenewsherald.com/news/2005-11-01/World/012.html, accessed 1/4/2011).

37. Ibid.

38. James Webb, *Born Fighting, How the Scots-Irish Shaped America* (Broadway Books, 2004).

39. A. A. Gray, "Camels in California," California Historical Society Quarterly, Vol. 9, No. 4 (California Historical Society, Dec. 1930): 299-317.

Chapter 9

40. William C. Barnard, "Hi Jolly and the US Camel Corps: Prospector Convinced Arizona Still Has Camels" (Associated Press, 1940s [Nov. 21; year unknown]) As told to William C. Barnard by prospector Bill Keiser. (http://www.scvhistory.com/scvhistory/hijolly-ap.htm, accessed 7/18/2014).

41. Campbell, *Wealth and Power in Antebellum Texas*.

42. Gonzales County, The Handbook of Texas Online, Texas State Historical Association TSHA, (http://www.tshaonline.org/handbook/online/articles/hcg07. The Handbook of Texas online, Texas State Historical Association TSHA. http://www.tshaonline.org/, accessed 3/28/14).

43. From Census records: Dennis Carroll c 1822 NC, unmarried; Lewis Carroll c 1825 NC, buried Gonzales Co. TX. Carroll Cemetery; unmarried; Mary A.

Carroll c 1827 NC, married Dr. Johnson; George Washington Carroll 28 Oct 1832 GA, married Nancy Abigail McElroy 14 Nov 1895, Buried Dayton Cemetery, Dayton WY; Elizabeth B Carroll c 1838 GA, married Dr James Cox; Jacob Carroll, Jr. c 1839 GA, married Prudence; Sarah "Sallie" Carroll 4 May 1840 GA, married James Turk 14 Sept 1861 Gonzales TX, died 20 Jan 1882 Gonzales TX, Carroll Cemetery; Columbus Carroll c 1843 GA.

Lois Turk Shrader reportedly inventoried the Carroll family cemetery in 1981. She discovered it 10.1 miles east on Hwy 90A Rt on FM 443 and .6 miles South to George Turk property on left. No gate barred the entrance when Lois explored and the site is approximately 400 yards from the road.

44. *The History of Gonzales County, Texas.*

45. "Antebellum Texas," The Handbook of Texas Online (https://www.tshaonline.org/handbook/online/articles/npa01, accessed 3/28/14). "Married women retained title to property such as land and slaves owned before they wed, had community rights to all property acquired during a marriage, and had full title to property that came into their hands after divorce or the death of a husband. These rights allowed Texas women to head families, own plantations, and manage estates in ways that were anything but passive and submissive."

46. Edward MacLysaght, *The Surnames Of Ireland* (Irish University Press, 1969). The name "Turk" comes either from MacTurk (Mac Tuirc), a Scottish name found in Antrim, Ireland, or more often from an abbreviated form of Turkington, An English toponymic closely associated with County Armagh, Ireland, since the 17th Century.

47. Paul Boethel, *The Free State of Lavaca* (Weddle Publication, 1977), Chapter 10, "When the Dust Settled on the Patriarch's House." (http://www.tamu.edu/faculty/ccbn/dewitt/smothers2.htm, accessed 3/28/14). Sons of DeWitt Colony Texas, http://www.tamu.edu/faculty/ccbn/dewitt/dewitt.htm. (Wallace L. McKeehan, ed.)

48. Ibid.

49. Ibid. Boethel begins chapter 10 with a quote from the State's Attorney for Lavaca County, who in 1891 presided over a murder trial and the defendants were Bird Kelly (John Smother Sr.'s son-in-law) and John [Wesley] Smothers (John Sr.s son.) The States Attorney lamented that "Bird Kelly and John [Wesley] Smothers, 'are members of large and influential families'; that there existed 'combinations and influences' which would preclude the State's getting a fair shake. [State of Texas vs. Bird Kelly et al., No. 2840, Criminal Docket, District Court, Lavaca County]".

50. Ibid. William Smothers to H. R. Runnels, May 5, 1858. Governor's Letters, Texas State Archives, Austin.

51. Ibid. Bill Smothers "served as commanding officer until May 7, 1863, when he resigned because of a back injury. He returned to Hallettsville and engaged in hauling freight and supplies out of Mexico. He was killed in a 'shoot-out' later in the year, reportedly by one named White, a Union agent. For posterity, he left a son, A.J. Smothers, sheriff of Lavaca County, 1882-1888, and a flock of other heirs, most all of whom were reputable citizens."

52. Ibid. By the mid 1880s, "Smothers the patriarch, or kingfish as he would be called today, was a man of means and a large family."

Chapter 10

53. Betty D. Fly and Craig H. Roell, "Fly, George Washington LaFayette," Handbook of Texas Online (Texas State Historical Association, uploaded on 6/12/2010; modified 1/18/2013; http://www.tshaonline.org/handbook/online/articles/ffl39, accessed 7/18/2014).

54. *The History of Gonzales County, Texas.*

55. Ibid.

56. Ibid.

57. Block, *A History of Jefferson County, Texas.* Chapter VII, "Early Transportation And Commerce" (http://www.wtblock.com/wtblockjr/History%20of%20Jefferson%20County/chapter%207.htm, accessed 7/18/2014).

58. Ibid.

59. Lynn Jones, "Crystals and Conjuring at the Charles Carroll House, Annapolis, Maryland" (The African Diaspora Archaeology Network, http://www.diaspora.illinois.edu/a-aanewsletter/newsletter27.html#anchor213532, accessed 7/18/2014).

"During excavation of the ground floor of the Charles Carroll House, a number of quartz crystals and associated artifacts were recovered. On the verbal authority of several of these scholars, the crystals were thought to have been used by slaves and to be related to African divination and conjuring practices (Dr. Frederick Lamp, Baltimore Museum of Art; Dr. Peter Mark, Wesleyan University; Dr. Robert Farris Thompson, Yale University, 1992, personal communication). The excavation at the Carroll House was conducted by Archaeology in Annapolis, a research project operated by the Department of Anthropology, University of Maryland, College Park, and by Historic Annapolis Foundation.

"The property had been purchased in the early eighteenth century by Charles Carroll the Settler, who at his death in 1720 owned over 47,000 acres of land in Maryland including one-quarter of the lots in the city of Annapolis. In 1721, Charles Carroll of Annapolis (the Settler's son) built the brick mansion, which still stands on the property. The most famous occupant was Charles Carroll of Carrollton, a signer of the Declaration of Independence, a Maryland State Senator, and a framer of the Maryland Constitution.

"Among the many artifacts recovered were several caches of quartz crystals and other objects. The largest groups of crystals, 12 in all, were found grouped together in an area about 6 inches in diameter (Figure 1). Found with them was a tiny faceted glass bead and a smooth black stone. This group of items was covered with a pearlware bowl of English manufacture turned upside down over them. The bowl is hand-painted pearlware, blue on white design; the design looks like a large asterisk or sun-burst."

Chapter 11

60. Merle Harris, genealogist; research from family records, grave sites, county records, and more.

61. The children born to George Washington Carroll (1832-1895) and Nancy Abigail McElroy (1837-1925): George Granville (1860-1946), Lillian (1865-1930), Nellie (1869-?), Fred (1873-1909), Solon (1874-1925). See this family genealogy in Appendix B.

62. Merle Harris genealogical research.

63. Sheldon F. Gauthier, "Range Lore," from *American Life Histories: Manuscripts from the Federal Writers' Project, 1936-1940 collection*, story of A.M. Garrett (Library of Congress, http://lcweb2.loc.gov/wpa/34061216.html, accessed 3/28/2014).

64. Boethel, *The Free State of Lavaca*, Chapter 10.

65. Miriam Forman-Brunell, ed., *Girlhood in America: An Encyclopedia* (ABC-CLIO, 2001), 73.

66. Children of John M. Kirby and Rachel Sells Kirby: James Kirby (1862-1938), George Kirby (1863-1946), John Kirby, (1869 – 1943), Orpha (1867-1890), William (1871-1964) Emma (1872-1959), Dora (1875-1963), Charles (1879-1862). See this family genealogy in Appendix B.

Chapter 12

67. Old West Tales, from the transcript of a radio broadcast on 2003-2004 WNOL Wells, Nevada that was posted on Nevada Online Encyclopedia, but later removed. Found online about 2011.

68. Ibid. Later the Goodnight cattle trail from Texas to Wyoming was established by Charles Goodnight and John Adair.

69. Dr. Leroy Vaughn, *Black People and Their Place in World History* (C.U.R.E. Publishing. 2002).

70. *The History of Gonzales County, Texas*.

71. A.W. Hunter, Camp Wood, Texas, "Was Among The First to Go Up The Trail," (Frontier Times Magazine, Vol. 6, No. 8, May 1929; *see also* http://frontiertimesmagazine.blogspot.com/2013/09/frontier-times-magazine-vol-6-no-8-may.html, accessed 3/29/2014).

72. Ibid.

73. Harold Carroll, "George Granville Carroll, Early Western Pioneer," (in Carroll family records).

Chapter 13

74. According to genealogist Merle Harris, the counterpane was made by Esther Mundell, Ella's grandmother, and bears the date of 1846. Myrtle Carroll had no daughters. "So on conferring with her three sisters, it was decided to place it in a museum in Billings, Montana where there is a large collection of heirlooms mostly from old pioneers of the West." In 2005, Lilas Cox, Alta Carroll's daughter, distributed the contents of the museum to family members, including the heirloom counterpane. There was also an old oak chest in Myrtle's family which was said to have been brought over on the Mayflower according to family tradition.

75. Dee Brown, *The American West* (USA Touchstone, 1995).

76. S. Omar Barker, "Code of the Cow Country" (Cowboy Miner Productions,

1998). The author copied this poem from a tacked-up 8x10 prominently displayed on the kitchen wall of Alta Carroll's daughter, Lilas Cox. Later research established that it was published in Barker's *Cowboy Poetry: Classic rhymes* and in *Songs of the Saddlemen* (Sage Books, 1954; http://www.cowboypoetry.com/sobarker.htm, accessed 3/29/2014).

Chapter 14

77. Alta Wallace Carroll, "Just Married" (1981). Alta wrote "Just Married" late in her life for the benefit of following generations. The pages were given to me by her daughter, Lilas Cox.

78. Bruce Brockett, *The Line Rider, Fenced Trails and Other Poems* (1948). In Utah State University Merrill-Cazier Library Special Collections & Archives, Fife Folklore Archives.

PART II, Chapter 15

79. We mustn't forget the Picts, who also inhabited the isles. Reputed to have "disappeared," they were instead absorbed into the Gaelic society just as the Norse who attacked and enslaved the Gaels later intermarried with them.

80. Noel Ignatiev, *How the Irish Became White* (Ruteledge, 1995). Ireland was governed under a series of codes called Penal Laws which regulated every aspect of life – civil, domestic, and spiritual – and imposed "a caste status out of which no catholic, no matter how wealthy, could escape. The racial and class hierarchy was enforced by the Dissenters....descendants of soldiers settled by Cromwell and Scots settled later in Ulster."

81. Ibid.

82. Christine Kinealy, *A Death-Dealing Famine, The Great Hunger in Ireland* (London: Pluto Press, 1997).

83. "Food Exports From Ireland During The Famine Years (1845-1849)" (http://www.usbornefamilytree.com/irishfoodexports.htm, accessed 3/29/2014.)

84. "Irish Potato Famine" (The History Place, http://www.historyplace.com/worldhistory/famine/, accessed 7/18/2014).

"In deciding their course of action during the Famine, British government officials and administrators rigidly adhered to the popular theory of the day, known as laissez-faire (meaning let it be), which advocated a hands-off policy in the belief that all problems would eventually be solved on their own through 'natural means.'

Great efforts were thus made to sidestep social problems and avoid any interference with private enterprise or the rights of property owners. Throughout the entire Famine period, the British government would never provide massive food aid to Ireland under the assumption that English landowners and private businesses would have been unfairly harmed by resulting food price fluctuations.

In adhering to laissez-faire, the British government also did not interfere with the English-controlled export business in Irish-grown grains. Throughout the Famine years, large quantities of native-grown wheat, barley, oats and oatmeal sailed out of places such as Limerick and Waterford for England, even though local Irish were dying of starvation. Irish farmers, desperate for cash, routinely sold the grain to the British in order to pay the rent on their farms and thus avoid eviction. On June 29, 1846, the resignation of British Prime Minister Sir Robert Peel was announced. Peel's Conservative government had fallen over political fallout from repeal of the Corn Laws which he had forced through Parliament. His departure paved the way for Charles Trevelyan to take full control of Famine policy under the new Liberal government. The Liberals, known as Whigs in those days, were led by Lord John Russell, and were big believers in the principle of laissez-faire.

Once he had firmly taken control, Trevelyan ordered the closing of the food depots in Ireland that had been selling Peel's Indian corn. He also rejected another boatload of Indian corn already headed for Ireland. His reasoning, as he explained in a letter, was to prevent the Irish from becoming "habitually dependent" on the British government. His openly stated desire was to make "Irish property support Irish poverty."

As a devout advocate of laissez-faire, Trevelyan also claimed that aiding the Irish brought "the risk of paralyzing all private enterprise." Thus he ruled out providing any more government food, despite early reports the potato blight had already been spotted amid the next harvest in the west of Ireland. Trevelyan believed Peel's policy of providing cheap Indian corn meal to the Irish had been a mistake because it undercut market prices and had discouraged private food dealers from importing the needed food. This year, the British government would do nothing. The food depots would be closed on schedule and the Irish fed via the free market, reducing their dependence on the government while at the same time maintaining the rights of private enterprise." Unfortunately, laissez faire is the prevalent economic policy as of the writing of this manuscript.

85. Kinealy, *A Death-Dealing Famine, The Great Hunger in Ireland*.
86. Ibid.
87. Ibid.
88. The term "Troubles" refers to a period of time between the early 1960's and the 1998 Good Friday Accord negotiated by US President Bill Clinton. Violence between the mostly Protestant "Unionists," who wished to remain united with Great Britain, and the mostly Catholic "Republicans," who wish to be part of the Republic of Ireland, continues to this day. When I visited Belfast in 2004 I stayed at the Europa Hotel, which has the reputation of being the most bombed hotel in Europe. It was devastated in a bombing in 1993. I took a black taxi tour in Belfast but when I called to make the arrangements I was told, "The gates are closed so we won't be able to go today. There's a march." The Protestant Orange were marching, still celebrating Cromwell, and the Catholic were mourning their dead and reacting in their own ways to keep the hate alive. The "gates" (military barricades) were closed, literally dividing the city in an effort to reduce violence.
89. It is because of the term "Irish Defectionist" that I deem James to have been Catholic. Of course others immigrated as well. Britain's "restrictive legislation, combined with a rapid growth in population and increasing competition for fertile land, contributed to a large scale exodus of Presbyterians from Ulster to

colonial America in the second half of the eighteenth century. Catholics also chose to emigrate, but in smaller numbers." Kinealy, *A Death-Dealing Famine, The Great Hunger in Ireland*.

90. Ignatiev, *How the Irish Became White*.

91. Ibid.

Chapter 16

92. Martin Wallace, *A Little History of Ireland* (Apple Tree, 1994).

93. Michael Pollock, "The Molly Maguires, Labor Heroes or Ethnic Terrorists" (2001; http://castle.eiu.edu/historia/archives/2002/Pollock.htm, accessed 3/29/2014). This paper was written for Dr. Nora Pat Small's Social History Class in the fall of 2001.

94. Philip Jenkins, *Terrorism as Heritage: How the Molly McGuires Became a Tourist Attraction* (Pennsylvania State University; http://www.personal.psu.edu/faculty/j/p/jpj1/molly.htm, accessed 7/18/2014).

95. James McCague, *The Second Rebellion: The Story of the New York City Draft Riots of 1863* (New York: Dial Press, 1968).

96. Pollock, "The Molly Maguires."

97. Blog, Sceala Clannad T.D. www.sceala.com. Date of Access: 3/29/2104.

98. Pollock, "The Molly Maguires."

99. Jenkins, *Terrorism as Heritage*.

100. Robert V. Bruce, *1877: Year of Violence* (Bobbs-Merrill Company Inc., 1959). Bruce describes a country entering the industrial age with one of "the largest and bloodiest general strikes in history. The labor unrest and resulting violence erupted from Baltimore to Chicago to St. Louis. What began as a strike by railroad workers escalated into a general civil war between capital and labor, inciting the first fears of 'Communism in America,' and put down by Federal troops with sword and cannon. Along with the draft riots of '63, the general strike of 1877 is a major reason why Americans see National Guard armories in their home towns."

101. Merle Harris genealogical research. A letter describes how Walter's wife and another man conspired to murder him.

102. Margaret and Jannet married in 1875, and Jane in 1877. Sons John, 22, and James, 21, stayed or returned later to Pennsylvania. Thomas is pictured in a family photo taken in 1914 at Fayette, Pennsylvania, with his brother James and sister Jane and their families.

103. Written on the opposite side of James Carroll's gravestone: "Christina Bennet, Wife of James Snedden, Died Nov. 22, 1890 aged 64 Years, Gone but not forgotten. Late wife of James Carrol." Mackey Cemetery, Harrison Township, Boone County, Iowa. Wives of James McCarroll included: Catheran Sharp and Christena Bennett who had once been married to a James Snedden.

Chapter 17

104. [Prepared by] the Sheridan Branch American Association of University Women, "Historical Sheridan" (The Sheridan Chamber of Commerce, Sheridan,

Wyoming, date unknown.)

105. Ibid.

106. Ibid.

107. Dee Brown, *Bury My Heart At Wounded Knee* (Holt Paperbacks, 2007).

108. Ibid.

109. Ibid.

110. Ibid.

111. Kenneth Meadows, *Earth Medicine* (Element Books Limited, Shaftesbury, Dorset, revised edition 1996).

112. Robert Bauval and Adrian Gilbert, *The Orian Mystery: Unlocking the Secrets of the Pyramids* (Crown Publishers, 1994).

Chapter 18

113. Of the nine children that Martha Carroll gave birth to, only five survived to adulthood: Myrtle L. Carroll, 11/8/1887 – ; Robert E. Carroll, 10/9/1889 – 7/7/1966; Arthur C. Carroll, 7/14/1901 – 6/13/1944; Margaret Ruth Carroll, 8/12/1893 – ; James W. Carroll, 7/12/1900 – 1960. See this family genealogy in Appendix B.

114. Francy Osgard, from unpublished Memory Book about her grandparents, Robert E. Carroll (Peter Carroll's son) and LaVina W. Carroll.

115. Gayla, Koerting, "Panic of 1873" (American Business, March 2, 2001). The Johnson County War might be seen as a Western response to the issues of 1873 that led Easterners and Midwesterners to riot. "The Panic of 1873 represented the first great crisis of industrial capitalism in the United States, and it altered the nature of economic enterprise, political ideology, and labor rights. The resulting depression caused widespread tension between laborers and capitalists, dividing the country along class lines. The Philadelphia banking firm of Jay Cooke and Company handled most of the government's loans during the Civil War. It began to invest heavily in the railroads September, 1873, the firm had overextended itself and declared bankruptcy. This unregulated, speculative credit created a vast overexpansion of the nation's railroad network. Paper money soon depreciated, and the impact fell on the domestic economy.

"The situation worsened when New York banks loaned money to railroads that expected to raise funds for repayment by selling bonds before the notes came due. There was no central national bank to shield the economy from the brunt of the railroads' collapse, so a chain reaction of bank failures resulted. The stock market plummeted, and the New York Stock Exchange was closed for ten days. Between 1873 and 1878, eighteen thousand businesses failed and the unemployment rate reached 14 percent.

"One-quarter of New York City's labor force was unable to find work in 1874. The panic caused companies to hoard cash receipts rather than depositing them in banks, so payrolls could not be met. In previous decades, workers had concentrated on such issues as greenbacks, cooperatives, and the eight hour workday. Now, they simply sought to maintain their pre depression wages or find unemployment relief. Some moved toward socialism.

"The emergence of the corporation in the United States coincided with the rise of the railroad industry and of organized labor movements to counter the influence of monopolistic practices. Political leaders were reluctant to involve the federal government too heavily in the private sector, but this mentality shifted during the late nineteenth century, when unions began asking the government to intercede on their behalf. Congress finally enacted the Interstate Commerce Act of 1887 to regulate railroads and the Sherman Antitrust Act of 1890 to prevent monopolistic companies from gaining total control in an industry. By the turn of the twentieth century, politicians and Progressive reformers created many of the United States' modern regulatory agencies; these departments included the Interstate Commerce Commission, the Food and Drug Administration, and the Federal Trade Commission."

116. Frank N. Schubert, "The Suggs Affray: The Black Cavalry in the Johnson County War" (Western Historical Quarterly Vol. 4, No. 1, January 1973), quoting Taylor Quintard, Jr., *In Search of the Racial Frontier*.

In 1890 small ranchers in Wyoming protested the territory's 1884 law that allowed the major cattle ranches to claim all unbranded cattle. Unable to get the law amended in a legislature dominated by powerful cattlemen, the small ranchers resorted to claiming unbranded cattle. The Wyoming Stock Growers' Association, which represented the large ranchers, immediately labeled its opponents "rustlers."

117. "Historical Sheridan."

118. *Lonesome Dove* website, "Today in Old West History, July 20" (http://www.knology.net/~lonesomedove/tiowhjul.html, accessed 3/29/2014).

119. David F. Norman, "Nate Champion and the Red Sash Gang" (1994; http://archive.is/52uXa, accessed 3/14/2014).

120. Wikipedia, "Johnson County War" (http://en.wikipedia.org/wiki/Johnson_County_War, accessed 7/19/14).

121. "Historical Sheridan."

122. Schubert, "The Suggs Affray."

123. Quintard Taylor, Jr., *In Search of the Racial Frontier* (W.W. Norton & Company, 1998), Chapter 6, "Buffalo Soldiers in the West, 1866-1917."

124. "Historical Sheridan."

Chapter 19

125. Nationwide women didn't have the right to vote until 1920 and the passage of the Nineteenth Amendment to the United States Constitution, which states: "The right of citizens of the United States to vote are not be denied or abridged by the United States or by any state on account of sex."

126. Katherine Arnold, *Sheridan Inn, "The House of 69 Gables"* (Sheridan County Chamber of Commerce, 1967).

127. Peter served as a member of the House of Representatives for the Republican Party again in 1927 through 1931, at the 19th, 20th, and 21st Sessions.

128. David Colman, *A History of the Labor Movement in the United States* (Iowa Federation of Labor AFL/CIO, November 2000).

129. Milton Friedman, "The Social Responsibility of Business is to Increase Its Profits" (New York Times Magazine, 9/13/1970).

130. Francy Osgard, from unpublished Memory Book.

131. Obituary, unknown newspaper (probably *Sheridan Press*): Mrs. Martha Atherton Carroll who died at her home on Coffeen Avenue at 8 o'clock Saturday morning after a lingering illness with cancer, was born in Syracuse, N.Y., on November 4, 1868. She removed to Iowa with her parents when 9 years of age. On March 25, 1886, she was married to Peter Carroll and removed with him to Montana fifteen years ago. After residing there for five years she came to Wyoming and settled in Sheridan County. She was the mother of nine children, six boys and three girls, five of whom are still living and survive her. Peter Carroll, (1860-1949), married Martha Atherton, (1866-1912).

132. Francy Osgard, from unpublished Memory Book. On the upper floor of the Carroll furniture store, Robert E. and wife LaVina started radio station KWYO.

133. Harold J. Carroll personal notes.

134. Wilfred Owen, Dulce Et Decorum Est (published posthumously in 1920). *Dulce et decorum est pro patria mori* is a line from the Roman lyrical poet Horace's Odes. The line can be roughly translated into English as: "It is sweet and fitting to die for one's country." (Wikipedia, http://en.wikipedia.org/wiki/Dulce_et_Decorum_est, accessed 7/19/2014.)

PART III, Chapter 20

135. Sparticus Educational website, "Hitler" (http://www.spartacus.schoolnet.co.uk/GERhitler.htm, accessed 7/19/2014).

136. The Holocaust Project, "Timebase 1940, September 1940" (http://archive.today/lvOjS, accessed 7/19/2014).

137. Ibid.

138. 101st Airborne website, "The 101st Airborne during World War II" (http://www.ww2-airborne.us/18corps/101abn/101_overview.html, accessed 7/19/2014).

139. Stephen Ambrose, *Band of Brothers E Company, 506th Regiment, 101st Airborne From Normandy to Hitler's Eagle's Nest* (Simon & Schuster, 1992).

140. Mark Israel, alt-usage-english website, "Cut the Mustard" (http://alt-usage-english.org/excerpts/fxcutthe.html, accessed 7/19/2014). "To cut the mustard": meaning "to achieve the required standard" was first used as an expression in a 1902 O. Henry story. It may come originally from a cowboy expression, "The proper mustard" meaning "the genuine thing" and the use of "cut" to denote rank (as in "a cut above") dating from the 18th century. Other theories are that it is a corruption of the military phrase "to pass muster" as in assembled troops passing inspection.

Chapter 21

141. Mildred Carroll, interviewed by Maureen Carroll, February 2007. Mildred Holcombe Carroll (born August 6, 1917); parents Riley Lee Holcombe and Grace Stewart. The ship her husband, Art Carroll, served on was the USS Illinois (BB-7) launched on 4 October 1898. Reclassified in 1941 and renamed Prairie State,

she served as a Naval Reserve Midshipmen Training School at New York during WWII.

Chapter 22

142. Wikipedia, "USS Sturgeon (SS-187)" (http://en.wikipedia.org/wiki/USS_Sturgeon_(SS-187), accessed 7/19/14).

143. Admiral Nimitz commanded U.S. forces during the Battle of Midway in 1942

144. U.S. Army, "Central Pacific 1941 – 1943" (U.S. Army Center of Military History, http://www.history.army.mil/brochures/72-4/72-4.HTM, accessed 7/19/2014).

145. Wikipedia, "USS Sturgeon (SS-187)."

146. Ibid.

147. Ibid.

148. Ibid.

149. "The Art of Pauline Jackson" Her puzzles are collectors' items. She also did book jackets and other art work.

Chapter 23

150. William H Greenhalgh, Jr., *A View Over the Next Hill*, (Air University Review, September-October 1973). "The outbreak of World War II found the reconnaissance forces of most nations obsolete and impoverished. Camera development had continued but slowly, principally for mapping purposes. The Army Air Corps had a few slow observation aircraft, designed primarily for artillery spotting and visual reconnaissance in support of ground units; but they were certain to fall victim to even the most obsolete enemy pursuit planes. The logical step was to modify either civilian aircraft or other types of military aircraft for reconnaissance, again a less than satisfactory solution."

151. Don Eddy "Heaven" (International News, Volume Five, Number 45, march 5th, 1942).

152. Wikipedia, "USS Sturgeon (SS-187)."

153. Naval History & Heritage, "Dictionary of American Naval Fighting Ships" (Department of the Navy, Naval Historical Center, http://www.history.navy.mil/danfs/ , accessed 7/19/14).

154. Ibid.

155. Wikipedia, "USS Sturgeon (SS-187)."

156. The Montevideo Maru Memorial Committee, November 2009. "The Tragedy of The Montevideo Maru. A Time for Recognition" (http://asopa.typepad.com/files/_submissionfinal.pdf, accessed 7/19/2014). The notes of Commander Wright.

157. Ibid.

158. Ibid.

159. Rod Miller, "The Montevideo Maru" (www.montevideomaru.info, accessed 7/19/2014. Copyright permission sought and received 2/14/2013).

160. Ibid.

161. The Montevideo Maru Memorial Committee, November 2009. "The Tragedy of The Montevideo Maru. A Time for Recognition."

162. I was unable to review Art or Chester's military records. I do not know if they were indeed onboard when the Montevideo Maru was destroyed.

163. Lt. Col. Phillip S. Gage, Jr., letter to Myrtle Carroll, dated 8, June, 1947. "P.S. My wife's sister, Patricia de Lorimier, was married last week. I mention this because Bob may have mentioned her to you. While when visiting us for two months at Pinehurst, NC, Bob seemed rather fond of her."

164. Sparticus Educational website, "Hitler."

Chapter 24

165. 101st Airborne website, "The 101st Airborne during World War II."

166. Ibid.

167. Stephen Ambrose, *Band of Brothers E Company, 506th Regiment, 101st Airborne From Normandy to Hitler's Eagle's Nest* (Simon & Schuster, 1992).

168. Harold J. Carroll personal notes.

169. Ibid.

170. I have been unable to determine the date the photo was taken or the purpose of the camera. The emblem, from about 1942, is in excellent condition and is pictured on page 233.

171. Omar Khayyam (Persian poet, 1048 – 1131), Rubaiyat.

172. Photo proof of Harold Carroll in paratroop gear preparing to jump with large camera. Photo taken about 1943.

173. Enlargement of photo proof of Harold Carroll.

174. Sparticus Educational website, "Hitler."

Chapter 25

175. Reasons to think H took the photographs: (1) he was a photographer for the military and was requisitioned for other photo assignments. (2) The original photos were in his possession. (3) Military records of his whereabouts show he left Ft. Benning, Georgia in March 1943 and is next stationed at Camp Taccoa, Georgia; there were other photo assignments taking him between bases.

Reasons to think H did not take the photographs: H could not have taken the photo if it was snapped in England on March 23, 1944 at an inspection visit by Prime Minister Winston Churchill, Supreme Allied Commander Dwight D. Eisenhower, U.S., First Army commander Omar Bradley, etc. The photos may have been taken in England even though Eisenhower is not represented because some officers appear to wear British uniforms. Additional research may reveal the actual origin of the photos.

176. 101st Airborne website, "The 101st Airborne during World War II."

177. Ibid.

178. Stephen Ambrose, *Band of Brothers*.

179. Wills, *Put on Your Boots and Parachutes!: Personal Stories of the Veterans of the United States 82nd Airborne Division from the Second World War* (1992).

180. Wikipedia, "USS Sturgeon (SS-187)" (http://en.wikipedia.org/wiki/USS_

Sturgeon_(SS-187), accessed: 7/19/14).

181. Ambrose, *Band of Brothers*, p. 54. I don't believe that the pictures described by Ambrose and taken in England in 1944 are the same Churchill pictures of which I am in possession. See Note 175.

182. *Britain's Homage To 28,000 American Dead* (London Times, 1952). The book is dedicated: In Memory Of Lt. Col. R. C. Carroll, U.S. Army. One copy of this book was presented to the family of each fallen soldier. The copy given to Myrtle Carroll remains in excellent condition.

183. Ibid.
184. Ibid.
185. Ibid.
186. Ibid.

Chapter 26

187. Deryk Wills, *Put on Your Boots and Parachutes!* Interview of unknown serviceman.
188. Harold J. Carroll personal notes.
189. Wills, *Put on Your Boots and Parachutes!*
190. Ibid.

Chapter 27

191. Letter to Harold from Derek Wills, not dated.
192. Ambrose, *Band of Brothers*.Pg. 60
193. The name that Harold affectionately called his unborn child, Patricia.
194. Merle Harris had three children, all girls: Marjorie Harris Carroll, (1924 – 1967); Lucile Harris Carr, (1922 – 2011); Lois Harris Daughtery (1921 – 2004(?).

Chapter 28

195. Auxilliary Territorial Services was the women's branch of the British Army during WWII.
196. Wills, *Put on Your Boots and Parachutes!*
197. Ibid.
198. John Driscoll, quoted in Francy Osgard's Memory Book about her grandparents, Robert E. Carroll and LaVina W. Carroll.

Chapter 29

199. Wills, *Put on Your Boots and Parachutes!*
200. H, quoted by Deryk Wills in his book *Put on Your Boots and Parachutes!: Personal Stories of the Veterans of the United States 82nd Airborne Division from the Second World War* (published 1992).
201. There was great confusion as to who took the pictures found in Quorn. Harold possessed copies of correspondence between Deryk Wills and Carl Clausen regarding Deryk's presumption that Stan Weinberg, a second lieutenant in B Co., took the pictures. Harold also possessed original photographs taken at the same time and place of the same people. They were likely taken on the same

camera. Harold told me that the pictures in the Leicester Mercury were taken by him, and that he had dropped them off for development but never returned to England to retrieve them. According to Carl Clawson, some of the photographs were taken by him and Stanley Weinberg. Clearly, the photographs of Harold were taken by another.

202. Deryk Wills, *Put on Your Boots and Parachutes!*

PART IV, Chapter 30

203. Harold Carroll, interviewed by staff writer Niki Hale (Red Rock News, Sedona, Arizona, 6/8/1994).
204. Ibid.
205. Ibid.
206. Ibid.
207. Harold J. Carroll personal notes.
208. *Utah Beach To Cherbourg 6-27 June 1944* (Center of Military History U.S. Army Washington, D.C., 1990).
209. Harold J. Carroll personal notes.
210. The Associated Press (clipping from unknown newspaper obituary: "Vandervoort, 75, died Sunday.")
211. The Associated Press (another clipping from unknown newspaper).
212. Henry Wadsworth Longfellow, "The Battle of Lovell's Pond," poem first published in the Portland Gazette (November 17, 1820).

Chapter 31

213. George E Koskimaki., *D-Day with the Screaming Eagles* (House of Print, 1970; Casemate, 2002).
214. Ibid.
215. Harold J. Carroll personal notes.
216. Deryk Wills, *Put on Your Boots and Parachutes!*

Chapter 32

217. Harold Carroll, interviewed by staff writer Niki Hale (Red Rock News, Sedona, Arizona, 6/8/1994).
218. Ibid.
219. Harold J. Carroll personal notes.
220. Arthur Carroll was only fifty-three when he died, but he'd spent his formative years in the mines and smoked heavily.
221. From an old, yellowed newspaper clipping. Research shows it was written by Reuben Goldsmith, *Gleams and Dreams: A Book of Verse* (J.T. White and Co., 1918).

Chapter 33

222. Harold Carroll, interviewed by Niki Hale.
223. Ibid.
224. Harold J. Carroll personal notes. "As to the actual location of Starvation

Maner/Hill I have been referring to a map: Le Havre 2nd Edition; Geographical Section, General Staff, No 4072, Published by War Office 1944, 2nd Edition 1942. Also a more detailed later map 'illes Anglo-Normandes: (Dressee par la Manufacture Francaise des Pneumatiques MICHELIN.'"

Chapter 34

225. Harold Carroll, interviewed by Niki Hale.

226. Brooks Kleber, former Nazi POW, interviewed by Alexander S. Cochran, Jr., "Trauma of Capture" (Military History, Vol. 1, No. 4, February 1985, pp 42-49, http://www.oflag64.us/Oflag64/The_Long_Cold_March_files/cochran_alexander-interview.pdf, accessed 7/19/2014).

227. Ibid.

Chapter 35

228. Harold Carroll, interviewed by Niki Hale.

229. Harold J. Carroll personal notes.

Chapter 36

"Kriegie Humor," by Lester H. Russell, page 327: Col. C. Ross Greening and Sgt. Angelo M. Spinelli, *The Yankee Kriegies: How Our POW's Made "Little Americas" Behind Nazi Barbed Wire* (prepared by National Council of Young Men's Christian Association; published by Guide Printing Co., NY). (Booklet in excellent condition.)

230. H interview/notes as published in Deryk Wills, *Put on Your Boots and Parachutes!*

231. Harold J. Carroll personal notes.

232. *Oflag 64 – the Fiftieth Anniversary Book*, Commissioned by the Anniversary Committee for the Kriegy Reunion, (Evanston Publishing, Inc., Evanston, Illinois, 1993). (Harold Carroll's copy remains in excellent condition.)

233. Kleber, former Nazi POW, interviewed by Alexander S. Cochran, Jr.

Chapter 37

234. *Oflag 64 – the Fiftieth Anniversary Book*, Commissioned by the Anniversary Committee for the Kriegy Reunion.

235. Ibid.

236. Ibid.

237. Harold J. Carroll personal notes. Clarence Meltesen asked H about the camp in a letter. Harold's response was published in Clarence R. Meltesen, *Roads To Liberation From Oflag 64*, (Oflag 64 Press, San Francisco, 1987, 2nd edition 1990). Harold Carroll's copy is autographed, "27 Aug 90 Clarence R Meltesen Let's keep in touch."

238. Ibid.

239. Thom Wilborn, On the Shortwaves.com, "Short-Wave Radio Monitors Let Families Know of Their Capture" (Originally appeared in Disabled American Veterans magazine, July/August 1998; http://www.ontheshortwaves.com/capture.html, accessed 7/19/2014).

240. Notice removed from Oflag 64, German prisoner of war camp, located in Poland, by Ralph Tedeschi after escaping from the Germans, January 23, 1945. Harold Carroll had a copy.

Chapter 38

241. Brooks Kleber, Former Nazi POW, Interviewed by Alexander S. Cochran, Jr.
242. Harold Carroll, interviewed by Niki Hale.
243. Harold J. Carroll personal notes.

Chapter 39

244. Harold Carroll, interviewed by Niki Hale.
245. *Oflag 64 – the Fiftieth Anniversary Book*, Commissioned by the Anniversary Committee for the Kriegy Reunion.

Chapter 40

246. Meltesen, *Roads To Liberation From Oflag 64*.

Chapter 41

247. Meltesen, *Roads To Liberation From Oflag 64*.
248. Harold J. Carroll personal notes, as told to Meltesen. *Roads To Liberation From Oflag 64*.

Chapter 42

249. Harold Carroll, interviewed by Niki Hale.
250. Sparticus Educational website, "Hitler." Controversy still surrounds the circumstances of Hitler's death. It is widely believed, however, that Hitler committed suicide by gunshot on April 30, 1945.

Chapter 43

251. Harold J. Carroll personal notes.

Chapter 44

252. *Britain's Homage To 28,000 American Dead* (London Times, 1952). The book is dedicated: "In Memory Of Lt. Col. R. C. Carroll, U.S. Army." One copy of this book was presented to the family of each fallen soldier. The copy given to Myrtle Carroll remains in excellent condition.
253. Letter to Harold Carroll from William Ekman, dated 15 October 1944.
254. Koskimaki., *D-Day with the Screaming Eagles*.
255. Researcher Deryk Wills, in a letter to Harold, said, "I was interested that your brother was in the 501st. Because there is a Lt. Col. Robert C. Carroll listed as KIA, and is credited with a silver star. There is also a mention of him in the book, 'Ridgway's Paratroopers' by Clay Blair on Page 240 as being the 1st Battalions' Commander attacking the Douve River above Carentan and being killed in the jump. In the history of the 101st, 'Rendezvous with Destiny', Page

111, describes that the 1st Battalion were badly scattered with only 18 of the 45 planes unloading over the DZ. Carroll is also mentioned again. There is also a reference in Devlin's 'Paratrooper' on page 400."

256. Bill Paul, "POW's: Four Decades of U.S. Abandonment" (Wall Street Journal, 8/13/1987, http://digitalcollections.library.cmu.edu/awweb/awarchive?type=file&item=414751, accessed 7/19/2014).

257. Forrest W. Howell, "Kidnapped American POW's," (reprinted in *The Veterans Voice and Senior Review*, October 1987.)

258. National Alliance for Families, "WWII" (www.nationalalliance.org/vietnam/ovrvw02.htm, accessed 7/19/2014).

259. Joseph Pearce, *C.S. Lewis and the Catholic Church* (Ignatius Press, 2003).

Previously Published:

Michel De Trez, in American Warriors Pictorial History of the American Paratroopers Prior to Normandy, reproduced photos of Harold Carroll & Stanley Weinberg and B Co. on page 56 and page 169, with permission from Deryk Wills and credited to Stanley Weinberg.

Clarence R Meltesen, Road to Oflag 64, ISBN 0-9627005-0-9 Copyright, 2d Ed 1990, P 175. Clarence reproduced a copy of Harold's "The Man Who Came to Dinner" program, published with permission. Harold Carroll is referenced on pages 86, 179, 234, 235, and 346. The book is signed, "27 Aug 90 Clarence R Meltesen Let's keep in touch."

BIBLIOGRAPHY

101st Airborne website. "The 101st Airborne during World War II." http://www.ww2-airborne.us/18corps/101abn/101_overview.html, accessed 7/19/2014.

Ambrose, Stephen. *Band of Brothers: E Company, 506th Regiment, 101st Airborne From Normandy to Hitler's Eagle's Nest.* New York: Simon & Schuster, 1992.

"Antebellum Texas." The Handbook of Texas Online. https://www.tshaonline.org/handbook/online/articles/npa01, accessed 3/28/14.

Arnold, Katherine. *Sheridan Inn, "The House of 69 Gables."* Sheridan County Chamber of Commerce, 1967.

Barker, S. Omar. *Cowboy Poetry: Classic Rhymes.* Cowboy Miner Productions, 1998. http://www.cowboypoetry.com/sobarker.htm, accessed 3/29/2014.

Barnard, William C. "Hi Jolly and the US Camel Corps." Associated Press, 1940's (date unknown). http://www.scvhistory.com/scvhistory/hijolly-ap.htm, accessed 3/29/2014.

"Big Horn Mountains Tourist and Recreation Guide." Supplement to the Sheridan Press, Sheridan, Wyoming, July 21, 1967.

Block, W.T. *A History of Jefferson County, Texas: From Wilderness To Reconstruction.* http://www.wtblock.com/wtblockjr/history%20of%20jefferson%20county/Introduction.htm, accessed 7/18/2014.

Boethel, Paul. *The Free State of Lavaca.* http://www.tamu.edu/faculty/ccbn/dewitt/smothers2.htm. Sons of DeWitt Colony Texas. http://www.tamu.edu/faculty/ccbn/dewitt/dewitt.htm, accessed 3/28/14, edited by Wallace L. McKeehan.

Britain's Homage To 28,000 American Dead, book published by the London Times, 1952.

Brockett, Bruce. *Fenced Trails and Other Poems.* 1948. Special Collections & Archives, Fife Folklore Archives, Utah State University Merrill-Cazier Library.

Brown, Dee. *The American West.* USA. Touchstone, 1995.

Brown, Dee. *Bury My Heart At Wounded Knee.* Holt Paperbacks, 2007.

Bruce, Robert V. *1877: Year of Violence.* Bobbs-Merrill Company Inc., 1959.

Brucer, Marshall, Lt. Col. (ed.) *A History of Airborne Command and Airborne Center.* "Published by the Command Club for its members." Galveston, Texas.

Campbell, Randolph B. *Wealth and Power in Antebellum Texas.* College Station, Texas A&M University Press, 1977.

Campbell, Vada. "Montgomery County, Our Heritage." Unpublished.

Carroll, Alta. "Just Married." Unpublished, 1980.

Carroll, Alta. "Kootnai Christmas." Unpublished, 1980.

Carroll, Harold J. Notes, WWII Era Letters, & Correspondence.
Notes about Marjory, travels through Nazi occupied Europe, and escapes from POW camps. Including notes written on maps.

WWII letters written between family members and others, and sent by the War Department in 1944 and 1945.

Handwritten and typed correspondence with M. Carroll, Derrik Wills, Clarence Meltesen, and others.

Colman, David. *A History of the Labor Movement in the United States.* Iowa Federation of Labor AFL/CIO, November 2000.

Cochran, Alexander S., Jr. "Trauma of Capture" (interview of Brooks Kleber, former Nazi POW). Military History, Vol. 1, No. 4, February 1985, pp 42-49, http://www.oflag64.us/Oflag64/The_Long_Cold_March_files/cochran_alexander-interview.pdf, accessed 7/19/2014.

"Daily News Digest." Daily War News Summary. Headquarters, Camp Mackall, North Carolina. August 7, 9, & 11, 1943.

Department of the Navy. *Dictionary of American Naval Fighting Ships, Naval History & Heritage.* Naval Historical Center. http://www.history.navy.mil/danfs/ accessed 3/28/14.

De Trez, Michel. American Warriors, Pictorial HIstory of the American Paratroopers Prior to Normandy. D-Day Publishing, 1994.

Eddy, Don. "Heaven." International News, Vol. Five, No. 45, March 5th, 1942.

Fly, George Washington Lafayette, The Handbook of Texas online, Texas State Historical Association TSHA, http://www.tshaonline.org/handbook/online/articles/ffl39, accessed 3/29/2014.

Food Exports From Ireland During The Famine Years (1845-1849). http://www.usbornefamilytree.com/irishfoodexports.htm, accessed 3/29/2014.

Ford, Tennessee Ernie. "The Ballad of Davy Crockett lyrics." http://www.songlyrics.com/tennessee-ernie-ford/the-ballad-of-davy-crockett-lyrics/ accessed 5/15/2014.

Forman-Brunell, Miriam, ed. *Girlhood in America: An Encyclopedia.* ABC CLIO, June 8, 2001.

Freedman, Milton. "The Social Responsibility of Business is to Increase Its Profits." New Yorker Times Magazine, September 13, 1970.

Garland C. Bagley. *History of Forsythe County GA, 1832-1932.* Boyd Publishing Company, 1997.

Gauthier, Sheldon F. "Range Lore," from *American Life Histories: Manuscripts from the Federal Writers' Project, 1936-1940 collection.* "Story of A.M. Garrett." Library of Congress, http://lcweb2.loc.gov/wpa/34061216.html, accessed 3/28/2014.

Gayla, Koerting. "Panic of 1873." American Business, March 2, 2001.

"Geronimo, 501st Parachute Infantry." Col. Howard R. Johnson, Commanding Officer. Vol. 1. Camp Mackall, N.C. Wednesday, May 26, 1943.

Goldsmith, Reuben. *Gleams and Dreams: A Book of Verse*. J.T. White and Co. 1918.

"Gonzales County." The Handbook of Texas Online. Texas State Historical Association TSHA. http://www.tshaonline.org/handbook/online/articles/hcg07. The Handbook of Texas online. Texas State Historical Association TSHA. http://www.tshaonline.org/, accessed 3/28/14.

Gray, A. A. *Camels in California*. California Historical Society Quarterly, Vol. 9, No. 4, Dec., 1930, pp. 299-317. Published by: California Historical Society.

Greenhalgh, William H., Jr. "A View Over the Next Hill." Air University Review, September-October 1973.

Hale, Niki. Red Rock News, Sedona, Arizona. 6/8/1994.

The Handbook of Texas Online, "The Regulator-Moderator War (1839 to 1844)." http://www.tshaonline.org/handbook/online/articles/FF/jgf1.html, accessed 2/1/2010.

[Prepared by] the Sheridan Branch American Association of University Women. "Historical Sheridan." The Sheridan Chamber of Commerce. Sheridan, Wyoming, date unknown.

"The History of Gonzales County, Texas." The Gonzales County Historical Commission, 1986.

The Holocaust Project. "Timebase 1940, September 1940." http://archive.today/lvOjS, accessed 7/19/2014.

Hunter, A. W. (of Camp Wood, Texas). "Was Among the First to Go Up the Trail." Frontier Times Magazine Vol 6, No. 8, May 1929; http://frontiertimesmagazine.blogspot.com/2013/09/frontier-times-magazine-vol-6-no-8-may.html, accessed 3/29/2014.

Ignatiev, Noel. *How the Irish Became White*. Ruteledge, 1995.

Israel, Mark. "Cut the Mustard." http://alt-usage-english.org/excerpts/fxcutthe.html. Date of access: 7/78/14.

Jenkins, Philip. "Terrorism as Heritage." Pennsylvania State University, http://www.personal.psu.edu/faculty/j/p/jpj1/molly.htm, accessed 3/29/2014.

"Johnson County War." http://en.wikipedia.org/wiki/Johnson_County_War, accessed 3/28/14.

Jones, Lynn. "Crystals and Conjuring at the Charles Carroll House, Annapolis, Maryland." University of Maryland. http://www.diaspora.illinois.edu/a-aanewsletter/newsletter.html#anchor213532. From The African Diaspora Archaeology Network, http://www.diaspora.illinois.edu/a-aanewsletter/newsletter27.html, accessed 3/29/2014.

Kinealy, Christine. *A Death-Dealing Famine, The Great Hunger in Ireland*. Pluto Press, London 1997.

Koskimaki, George E. *D-Day with the Screaming Eagles*. House of Print, 1970. Casemate, 2002.

Lofaro, Michael A. "Crockett, David." Handbook of Texas Online, Texas State Historical Association, uploaded 6/12/2010, modified 1/27/2014; http://www.tshaonline.org/handbook/online/articles/fcr24, accessed 7/18/2014.

McCague, James. *The Second Rebellion: The Story of the New York City Draft Riots of 1863*. New York, Dial Press, 1968.

MacLysaght, Edward. *The Surnames Of Ireland*. Shannon, Ireland, Irish University Press 1969

McKaughan, Joshua Lee. "'Barcelonia' Neckerchiefs, Teaware, and China Plates: Kinship Status and the Division of Fourth Creek Church." Journal of Backcountry Studies, Vol. 1 No. 3, Spring 2008; http://www.partnershipsjournal.org/index.php/jbc/article/view/34/23, accessed 7/21/2014.

McKeehan, Wallace L., ed. "DeWitt Colony Papers Documents and Correspondence 1824-1826." http://www.tamu.edu/faculty/ccbn/dewitt/papers24-26.htm. Sons of DeWitt Colony Texas, http://www.tamu.edu/faculty/ccbn/dewitt/dewitt.htm, accessed 3/28/14.

Meadows, Kenneth. *Earth Medicine*. Element Books Limited, Shaftesbury, Dorset, revised edition 1996.

Meltesen, Clarence R. *Roads To Liberation From Oflag 64*. 1987, 2nd edition 1990.

Miller, Rod. "The Montevideo Maru." www.montevideomaru.info, accessed 3/28/14. Copyright permission sought and received Date: 2/14/2013.

The Montevideo Maru Memorial Committee, November 2009. "The Tragedy of The Montevideo Maru. A Time for Recognition." http://asopa.typepad.com/files/_submissionfinal.pdf, accessed 7/19/2014.

Mt. Tabor Baptist Church. 1800's Membership Lists of Mt. Tabor Baptist Church, GA.

Norman, David F. "Nate Champion and the Red Sash Gang." 1994; http://archive.is/52uXa, accessed 3/14/2014.

Nuttall, Thomas. *A Journal of Travels into the Arkansas Territory During the Year 1819*. University of Oklahoma Press, 1980.

Oflag 64 – the Fiftieth Anniversary Book. Commissioned by the Anniversary Committee for the Kriegy Reunion. Evanston Publishing, Inc., Evanston, Illinois. 1993. (Harold Carroll's copy remains in excellent condition.)

"Old West Tales." 2003-2004 WNOL Wells, Nevada ON LINE! Date of Access: 2/1/2010.

Osgard, Tracy. from Memory Book about her grandparents, Robert E. Carroll and LaVina W. Carroll.

Owen, Wilfred. "Dulce Et Decorum Est" was published posthumously in 1920.

Pollock, Michael. "The Molly Maguires, Labor Heroes or Ethnic Terrorists."

430 | BIBLIOGRAPHY

This paper was written for Dr. Nora Pat Small's Social History Class in the fall of 2001. http://castle.eiu.edu/historia/archives/2002/Pollock.htm, accessed 3/29/2014.

Prisoners of War Bulletin. Vol. 3, Nos. 3, 4, & 5, March, April, & May 1945. American National Red Cross.

Program & Menu. "Christmas 1944." OFLAG 64, December 1944.

Program for "The Man Who Came To Dinner." Little Theater, OFLAG 64, October 11 – 15 1944.

Sceala Clannad T.D. www.sceala.com. Date of Access: 3/29/2104.

Schubert, Frank N. "The Suggs Affray: The Black Cavalry in the Johnson County War," Western Historical Quarterly Vol. 4 No. 1, January 1973.

"The Secret Custer Letter." The Denver Post, September 14, 1975.

Service, Robert. "The Cremation of Sam McGee." 1907.

"The Sheridan Press Wrapper Spring 1987." Fort Phil Kearney, Sheridan Press, November 2 1886.

"The Sheridan Story 100 years 1882 – 1982." Sheridan Press, 1982.

Shipman, Daniel. (1801-1881), *Frontier Life: 56 Years in Texas in 1879*. Texas State Historical Association, http://www.tshaonline.org/handbook/online/articles/fsh30. Date of Access: 3/29/2014.

Slatta, Richard W. *The Cowboy Encyclopedia*. http://social.chass.ncsu.edu/slatta/essays/blackcowboys.htm, accessed 2/1/2010.

Sparticus Educational website. "Hitler." http://www.spartacus.schoolnet.co.uk/GERhitler.htm, accessed 7/19/2014.

Stanton, Bruce. *The Regulator Moderator War*. http://www.kilgorenewsherald.com/news/2005-11-01/World/012.html, accesed 1/4/2011.

Starr, Emmet. *History of the Cherokee Indians and their legends and folk lore*. The Warden Co., 1922.

Stewart, Richard W., ed. *American Military History*, Vol. II, Ch. 2. Center of Military History, United States Army, Washington, DC, http://www.history.army.mil/books/AMH-V2/AMH%20V2/, accessed 7/8/14.

"Tale of a City." U.S. Office of War Information, 1942.

"Taylor, Creed." The Handbook of Texas online. Texas State Historical Association, http://www.tshaonline.org/handbook/online/articles/fta17, accessed 3/29/2014.

"Today in Old West History, July 20." *Lonesome Dove* website, http://www.knology.net/~lonesomedove/tiowhjul.html, accessed 3/29/2014.

U.S. Army. "Central Pacific 1941 – 1943." U.S. Army Center of Military History, http://www.history.army.mil/brochures/72-4/72-4.HTM, accessed 7/19/2014.

"Utah Beach To Cherbourg 6-27 June 1944." Center of Military History U.S.

Army Washington, D.C., 1990.

Vaughn, Leroy. *Black People and Their Place in World History.* 2002.

Wallace, Martin. *A Little History of Ireland.* Apple Tree, 1994.

Wikipedia. "USS Sturgeon (SS-187)." http://en.wikipedia.org/wiki/USS_Sturgeon_(SS-187), accessed 7/19/14.

Webb, James. *Born Fighting, How the Scots-Irish Shaped America.* Broadway Books, New York, 2004.

Wikipedia. "Women's Suffrage in the US." http://en.wikipedia.org/wiki/Women's_suffrage_in_the_United_States, accessed 3/28/14.

Wilborn, Thom. On the Shortwaves.com. "Short-Wave Radio Monitors Let Families Know of Their Capture." (Originally appeared in Disabled American Veterans magazine. July/August 1998.) http://www.ontheshortwaves.com/capture.html, accessed 7/19/2014.

Wills, Deryk. *Put on Your Boots and Parachutes!: Personal Stories of the Veterans of the United States 82nd Airborne Division from the Second World War.* Derek Wills, 1992.

www.ingramcontent.com/pod-product-compliance
Lightning Source LLC
Chambersburg PA
CBHW081332080526
44588CB00017B/2594